Elegant Arches, Soaring Spans

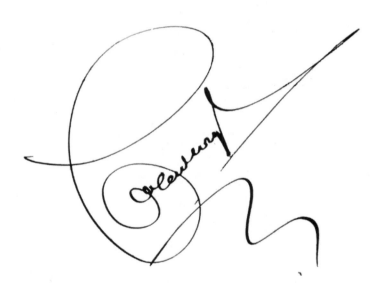

Elegant Arches, Soaring Spans

꙰

C. B. McCullough
Oregon's Master Bridge Builder

by Robert W. Hadlow

Oregon State University Press
Corvallis

The glossary in this book is excerpted from *The Portland Bridge Book*, 2nd edition, revised and expanded (Oregon Historical Society Press, 2001), and was written by Sharon Wood Wortman and Ed Wortman. It appears with permission of the book's author, Sharon Wood Wortman.

Photograph of Yaquina Bay Bridge on the front cover,
courtesy Jerry Robertson, Oregon Department of Transportation.
Photograph of C. B. McCullough, on the Isaac Lee Patterson Bridge, 1932,
frontispiece and on the front cover, from the John P. McCullough Collection.

The paper in this book meets the guidelines for permanence and durability of the Committee on Production Guidelines for Book Longevity of the Council on Library Resources and the minimum requirements of the American National Standard for Permanence of Paper for Printed Library Materials Z39.48-1984.

Library of Congress Cataloging-in-Publication Data
Hadlow, Robert W. (Robert William), 1961–
  Elegant arches, soaring spans : C.B. McCullough, Oregon's master bridge builder / Robert W. Hadlow.— 1st ed.
      p. cm.
Includes bibliographical references and index.
  ISBN 0-87071-534-8 (alk. paper)
  1. McCullough, Conde B. (Conde Balcom), 1887-1946. 2. Highway engineers—Oregon—Biography. I. Title.
  TE140.M36 H33 2001
  624'.2'092--dc21

                                    2001002156

Oregon State University Press
101 Waldo Hall Corvallis OR 97331-6407
541-737-3166 • fax 541-737-3170
http://osu.orst.edu/dept/press

OREGON STATE
UNIVERSITY

❧ Dedication ❧

For my father

Contents

List of Illustrations

Preface & Acknowledgements

I began researching the life and accomplishments of C. B. McCullough during the summer of 1990, when I was lucky enough to find employment with the Historic American Engineering Record of the National Park Service on its twelve-week Oregon Historic Bridges Recording Project. I studied the construction history of a handful of McCullough bridges. More importantly, though, I was assigned the task of writing a short biography of McCullough. Over the years, this work grew to a Ph.D. dissertation, and now a book.

For their deep commitment to my dissertation project, I owe a great debt to my committee, Orlan J. Svingen, David H. Stratton, and Jerry B. Gough. Professor Svingen's editorial expertise is unequaled. Professor Stratton was a welcomed advisor on both my M.A. and Ph.D. programs. There is no match for Professor Gough's command of the history of science and technology.

The Margaret Pettyjohn Endowment, Department of History, Washington State University, and the WSU Summer Graduate Student Research Fellowship Program provided generous support. More recently, Dwight A. Smith and James C. Howland helped to advance the manuscript from dissertation to book.

Many contributed greatly to this project and I think of them as mentors and friends. Foremost are Dwight A. Smith, formerly cultural resources specialist with the Oregon Department of Transportation (ODOT), and Eric N. DeLony, chief and principal architect of the Historic American Engineering Record—I am now what DeLony likes to call a "pontist," or one who has a genuine interest in bridges. Elisabeth W. Potter, formerly National Register of Historic Places Nominations Coordinator with the Oregon State Historic Preservation Office, guided me with her unparalleled knowledge of Oregon's historic structures.

Much appreciation goes to Emory L. Kemp, former director of the Institute for the History of Technology and Industrial Archeology, West Virginia University, Morgantown, who is a long-time champion of research of the country's industrial and technological past. Terry Shike, former Oregon state bridge engineer, and his knowledgeable staff supported the 1990 Oregon Historic Bridges Recording Project. The ODOT Bridge Section continues to show a genuine interest in Oregon's bridge-building history and is striving to find new technologies to preserve McCullough's structures for future generations.

I will never forget my conversations with the late Ivan Merchant, retired Oregon state bridge engineer, whom C. B. McCullough hired fresh from college in 1929. His insights into McCullough's days as state bridge engineer were invaluable.

Several archival specialists, librarians, and curators provided generous assistance. Elizabeth Nielsen, Oregon State University Archives, Corvallis; and Betty Erickson, Iowa State University Archives and Special Collections, Ames, made available their collections for research. Michael McQuaide and colleagues at the Oregon State Archives, Salem, assisted me through four months of intensive research in Oregon State Highway Division records in early 1992. Alden Moberg, former librarian and Oregoniana specialist, Oregon State Library, Salem, opened his invaluable collections to me. Carolyn Philp and Jerry Robertson assisted me with ODOT's general files and historic photograph collection.

I thank James B. Norman and Jerry Robertson for use of their wonderful photographs.

I thank Mrs. John R. McCullough for making available her collection of McCullough papers, and John P. McCullough who never met his grandfather, C. B. McCullough, but now has a more complete idea of who and what he was.

Finally, I dedicate this book to my father, William J. Hadlow, D.V.M., D.Sc., a researcher who made great strides in veterinary pathology. He inspired me to pursue a research vocation.

Abbreviations

American Association of State Highway Officials (AASHO)
Highway Research Board (HRB)
Iowa State College (ISC)
Iowa State Highway Commission (ISHC)
League of American Wheelmen (LAW)
National League for Good Roads (NLGR)
Office of Public Road Inquiry (OPRI)
Office of Public Roads (OPR)
Office of Road Inquiry (ORI)
Oregon Agricultural College (OAC)
Oregon Department of Transportation (ODOT)
Oregon State College (OSC)
Oregon State Highway Commission (OSHC)
Oregon State Highway Department (OSHD)
Public Works Administration (PWA)
Reconstruction Finance Corporation (RFC)
Works Progress Administration (WPA)
U.S. Bureau of Public Roads (BPR)

❧ 1 ❧

Introduction

I N 1937, WHEN CONDE B. MCCULLOUGH was at the height of his career, a close friend asked him to describe his impressions of the engineer's role in society. McCullough replied, "From the dawn of civilization up to the present, engineers have been busily engaged in ruining this fair earth and taking all the romance out of it. They have cluttered up the landscape with hideous little buildings and ugly railroads." Modestly, McCullough hoped that his life work would bring back a small part of the romance lost.[1]

McCullough was one of a new breed of college-educated civil engineers. He was a pioneer in the movement to create a well-planned American highway system by researching road and bridge design from a scientific as well as practical perspective. Beginning in the early 1900s, McCullough argued that bridges should be built efficiently, economically, and attractively. He took his message to Oregon in the 1920s and 1930s and transformed the state's bridge building into a regional, national, and increasingly international showcase. The periodical *ENR* recognized McCullough in 1999 as one of the top ten bridge engineers among its list of engineering greats.

❧ ❧

IRON AND STEEL DOMINATED American bridge building in the late nineteenth century. They were the materials of choice among numerous private companies that marketed their bridges to local governments. These "catalogue bridge entrepreneurs" promoted metal truss types that each claimed were superior to the products of their competitors. Large networks of salesmen from enterprises such as the Phoenix Bridge

Company of Phoenixville, Pennsylvania, the King Bridge Company of Cleveland, and Andrew Carnegie's Keystone Bridge Company of Pittsburgh persuaded their clients with an expansive sales pitch rather than scientific fact. Bridges selected by county commissioners were supposed to provide years of service, but often did not. "The constant pressure to outsell and out produce competitors," David Plowden writes, "led to shoddy practices, carelessness, and sometimes deliberate dishonesty."[2]

This system ended around the turn of the twentieth century for two reasons. First, railroad companies, and later state highway departments, required bridge companies to bid competitively on projects and college-trained engineers scrutinized their proposals. Second, reinforced concrete emerged as the preferred construction material for short- and medium-length bridges.[3]

Many states had competent bridge engineering staffs, who usually created relatively short-span structures based on pedantic, unimaginative, pattern-book designs. But when larger and more elaborate structures were considered, states tended to contract with private consultants thought to be better suited for complicated technical designs. But stream crossings in the western states routinely required one-of-a-kind bridges. The highway bridge departments in these states, far from East Coast consulting firms, had to rely on the abilities of their own staffs. Many created standardized designs that were routinely modified for each project. While efficient and economical, these bridges lacked aesthetic qualities. In Oregon, state-sponsored bridge construction directed by Conde B. McCullough took a new and innovative direction with highway structures that were designed to blend with their natural setting.

McCullough designed efficient, economical, and elegant stream crossings. He was an original thinker with superior mathematical abilities, an eye for design, a great knowledge of bridge building worldwide, and efficient managerial skills. Like nineteenth-century engineering figures such as John Roebling and James Eads and their twentieth-century European counterparts Robert Maillart and Eugène Freyssinet, McCullough combined an engineer's desire for creating efficient structures with public funds and an architect's drive for aesthetic excellence.[4]

☺☺

FROM 1906 TO 1910, McCULLOUGH studied civil engineering at Iowa State College under the respected and progressive educator Anson Marston. Marston trained his students in a curriculum that emphasized the applied aspects of the field as well as its philosophical underpinnings. Technical ability combined with a well-rounded education, Marston believed, provided the foundation for creating sensible highway bridge designs for local, state, and national interests. McCullough's training under Marston and Professor John E. Kirkham, a respected civil engineer, prepared him to become one of the nation's most accomplished bridge designers. He learned from his mentors that experts should give unselfish service to society.[5]

In 1910–11, McCullough worked for the Marsh Engineering Company of Des Moines, where he learned firsthand the intense rivalry among private companies to win bridge-building contracts from county governments. James Marsh promoted his reinforced-concrete arch structures as superior to his competitors' designs, thereby influencing McCullough's future artistic ambitions.

From 1911 to 1916, McCullough worked for the Iowa State Highway Commission (ISHC) as its bridge engineer and assistant highway engineer. He was a member of a team of talented young college graduates serving under Thomas H. MacDonald, who later advanced to become chief of the U.S. Bureau of Public Roads (BPR), determining federal highway policy for nearly four decades. McCullough and his colleagues instituted a forward-thinking, efficient, and economical highway building program for Iowa, far advanced over the piecemeal, ineffective, and wasteful county-based approach of the past.

McCullough had a good opportunity to hone his skills as a bridge designer while he worked at the ISHC. Because there was little professional literature, he devoted long hours to basic and applied research in structural challenges, hoping to find efficient and economical solutions to vexing engineering problems. He excelled in the mathematics involved in engineering calculations, and became known as an authority on bridge engineering when he offered expert testimony in court cases against unscrupulous bridge entrepreneurs. The expertise he gained helped him rise above his counterparts scattered across the country in other fledgling state highway departments and earned him widespread professional recognition.

With that national reputation, McCullough was invited to Oregon to teach structural engineering at Oregon Agricultural College (OAC), the state's land-grant school (now Oregon State University). He took an enthusiastic interest in the evolution, seeing certain parallels with his Iowa experience. Federal participation in nationwide road improvement emphasized establishing state highway programs with highly qualified staffs. Accordingly, in 1919, the Oregon State Highway Commission (OSHC) lured McCullough from OAC to become its bridge engineer. This marked the beginning of a solid career in building efficient, economical, and often beautiful bridges from a state agency base.

Progressive Oregon hired McCullough to help create an ordered system of highway bridges. Surrounding himself with bright assistants—his former students from OAC and his friends and classmates from Iowa—he formed a close-knit organization of professionals. As an extraordinary managerial wizard, McCullough excelled as the director of a bridge department, creating low-cost custom-designed structures characterized by architectural elegance. He preached that designers, in constructing long-lasting highway bridges, must take into account geological considerations, roadway alignments, traffic density, load requirements, and scenic value.

Reinforced concrete became McCullough's preferred medium over steel or wood. Some experts at that time viewed concrete as an expensive material requiring tedious mathematical calculations, but McCullough argued that lower maintenance expenses offset initial costs. He preferred the reinforced-concrete arch in large-span construction and he believed the arch cast in concrete was more economical than obvious alternatives. The reinforced-concrete arches in McCullough's larger bridges built from 1919 to 1936 combine aesthetics with efficiency and economy. His vision became a key component in transforming Oregon into a traveler's paradise that would generate state revenues through increased fuel taxes and stimulate local markets.

Throughout the 1920s and 1930s, McCullough became a prolific author of books and articles for engineering periodicals and bulletins for the BPR. He also studied law, and his interest in the convergence of law and engineering prompted him to earn a law degree. In the mid-1940s, he wrote a two-volume text on this subject.

Great challenges were McCullough's lifeblood. His large-span designs throughout the state, especially on the Oregon Coast Highway (U.S. 101), launched him into national and international prominence. The Isaac Lee Patterson Bridge at Gold Beach (the Rogue River Bridge) was the first structure completed in the United States using French engineer Eugène Freyssinet's methods for reinforced-concrete arch rib precompression. McCullough's five large-span structures at Newport,

Waldport, Florence, Reedsport, and North Bend on the Oregon Coast Highway together cost $5.6 million and represented the pinnacle of bridge design for their use of engineering advances and their architectural style to fit the natural setting. Contemporary and subsequent critics saw them as masterpieces in design.

McCullough's accomplishments as Oregon's bridge engineer made him the BPR's logical choice to create several bridges for the Inter-American Highway. Between 1935 and 1937, he designed three short-span suspension bridges and a dozen other structures for the route through Panama, Honduras, Guatemala, and other Central American republics. When McCullough returned to Oregon in 1937, he promoted the suspension bridge as a realistic alternative to other design types because it fit well with his three-part design philosophy: economy, efficiency, and aesthetics.

Promoted to assistant state highway engineer in 1937, McCullough became an unwilling administrator removed from his preferred vocation. He turned his energy to researching highway department management questions and writing about engineering law.[6]

Many believe that, but for his untimely death in 1946, more spans with his signature elements might have been constructed. Nevertheless, McCullough had made his mark on Oregon; and most of the bridges for which he is best known are still an important part of the state's highway system. Employing technologies only developed in the past few decades to arrest and prevent degradation of steel and concrete construction, today's engineers are ensuring that McCullough's bridges will continue to be part of Oregon's landscape for decades to come.

❧ 2 ❧

Civil Engineering with Anson Marston

Personally, I am all for the old covered bridge. It wasn't wide enough or strong enough to carry the loads, it is true, but it was long enough to reach from one bank of the river to the other, which is all that a bridge is supposed to do anyway.[1]

Conde B. McCullough penned this in 1937 in response to a friend's question about his views on bridge types. He was fifty years old and had been an engineer for nearly three decades.

MᶜCULLOUGH'S YEARS IN IOWA, where he was raised, educated, and married, had great influence on his subsequent career choices. More than anything, enrolling at Iowa State College (ISC) determined the career path that he would follow.

Iowa State imparted a close student–mentor relationship. McCullough and his classmates studied under Anson Marston, who instilled in them high standards for their professional careers. Creating efficient structures, as all engineers were trained to do, was not enough for Marston. Instead, sensible designs should reflect serious thought on the economics of construction and, where practical, the aesthetics of form. McCullough incorporated these ideas into every project that he completed. They became the structural underpinnings of his philosophy, which vaulted him into becoming one of the leading bridge designers of the first half of the twentieth century.

It was not just that Anson Marston was McCullough's mentor that made their association so important; the formation of Marston's philosophy of engineering education in the nineteenth century and its role in twentieth-century progressive ideals are equally important. These qualities helped Marston and his student play an important role in the evolution of civil engineering in the United States. They created the backdrop for McCullough's emergence as a key participant in a period of nationwide road and bridge building.

In his first professional job, with the Iowa State Highway Commission, McCullough helped create highway bridges that were products of his own and his mentor's engineering philosophy. His bridge-building training combined farsighted thought in engineering, economics, and aesthetics—design qualities that he applied during his thirty-five-year career in bridge design and construction.

❧ ❧

CONDE BALCOM MCCULLOUGH's paternal great-grandparents, William and Mary McCullough, and their seven-year-old son, Boyd, were Scots-Irish people who left their native Ireland in 1832 for a new life in western Pennsylvania's Jefferson County. By 1848, Boyd McCullough had earned a degree in classics from Duquesne College, in Pittsburgh. He then enrolled in a seminary in Cincinnati, Ohio, and was subsequently ordained in the Reformed Presbyterian Church. In 1855, Boyd left Ohio for a pastorship in Detroit, Michigan, with his wife, Julia Ann, whom he had married while still at the seminary, and their three-year-old son, John Black McCullough. In 1872, Boyd resigned from the organized ministry and devoted the next four years to lecturing and preaching in Great Britain. Once back in the United States, he became affiliated with the United Presbyterian Church in 1875, and was assigned to a congregation in Pepin, Wisconsin.[2]

John Black McCullough studied medicine at the University of Michigan. He later met and married Lenna Leota Balcom, who had attended the State Normal School in Winona, Minnesota. In the early 1880s, they moved west to Dakota Territory, where John set up a medical practice.[3]

On 30 May 1887, Lenna McCullough bore a son, Conde Balcom, near Redfield, in Spinks County, in present-day South Dakota. By the 1890s, John B. McCullough and his family had relocated nearer relatives at Fort Dodge, Iowa, where John studied to become a missionary for the Disciples of Christ. He planned to minister to the sick and infirm through the church's Christian Missionary Society. Lenna McCullough cared for

Conde Balcom McCullough, c. 1900.
(John P. McCullough Collection)

Conde, her only child, and devoted time to the Woman's Relief Corps, a patriotic organization devoted to Civil War veterans.[4]

The comfortable life of the McCullough family ended abruptly when John fell and injured his spine. The accident left him bedridden. Because his mother had no way to earn a living, twelve-year-old Conde worked at odd jobs to help support his family. On 25 October 1904, John suddenly died; his attack was later attributed to a cerebral hemorrhage. He was fifty-three. Conde, by then a high school senior, continued working at part-time jobs while caring for his mother. He received his high school diploma the following spring.[5]

In October 1905, McCullough was eighteen years old; he weighed 135-pounds and was five feet six inches tall. He found employment with the Illinois Central Railroad in Cherokee, Iowa, working on a section gang that maintained portions of the Illinois Central's track. He gained important practical experience by helping determine topographic variations crucial in mapping rail lines. Serving as a surveyor's assistant, McCullough handled a chain used to measure distance and held a pole marked with graduations that another assistant read from a distance through a transit or telescope instrument. By the fall of 1906, this recent experience helped persuade him to enter the civil engineering program at ISC in Ames, seventy miles from Fort Dodge.[6] McCullough probably chose to attend Iowa State College because, as a resident of Iowa, he could enroll at the state's land-grant college tuition-free.

In 1906, ISC had an enrollment of eight hundred students. McCullough was one of sixty who specialized in engineering. He began his freshman year by studying mechanical drawing and surveying, mechanics of engineering and astronomy, and mathematics and physics. He also enrolled in college-wide required courses in the arts and foreign languages.

A faculty of four professors and instructors led by Anson Marston taught the engineering courses. McCullough's civil engineering curriculum had been created by Marston in the early 1890s and molded and reshaped in the ensuing fourteen years. Marston had designed the course to reflect his perspective on the future of engineering. He stressed highway and sanitary engineering and gave less consideration to railroads. Inherent in his philosophy was that students should focus on engineering, but at the same time take courses in mathematics, science, the arts, and languages.[7]

Marston had been dean of the School of Engineering for only two years when McCullough enrolled there in 1906. Since becoming a faculty member in 1892, he had helped to create a reputable program leading toward a bachelor of science degree in civil engineering. He believed that students should receive instruction in the "principles underlying the actual practice of each special branch of the profession," including the design and construction of railroad grades and bridges, sanitary sewers and municipal water systems, and hydraulics, or the water retention properties of soils. His students also briefly examined the design and construction of buildings, roofs, masonry structures, and highway bridges. Marston and his faculty directed seniors to focus a significant portion of their studies on one of three specialty fields—structural, railway, or hydraulic engineering.[8]

Marston became one of the foremost American engineering educators of his time because of his commitment to a curriculum that prepared

Anson Marston, 1913.
(Iowa State University Library/University Archives)

students for employment through practical education. He had formed his perspective on the structure of education in Estévan Antonio Fuertes' civil engineering course at Cornell University, in Ithaca, New York, where he had enrolled in 1885.[9]

The Morrill Act of 1862 had mandated American land-grant colleges and universities to offer schooling in agriculture and mechanic arts. Land-grant institutions sought to train students in a vocation while providing a balanced general education. Ezra Cornell, the patron of the school bearing his name, and others hoped that land-grant institutions would become integral parts of state systems of higher education. Cornell held a broad view of higher education for the masses and sought to found a university based on this principle.[10]

Cornell created its field of study in civil engineering when the university was founded in 1866. In 1873, the university hired a well-known engineer, Estévan Fuertes, who strengthened the program. Fuertes was born in Puerto Rico and educated in Spain. He completed graduate work in civil engineering at the Rensselaer Polytechnic Institute, in Troy, New York, in 1861. Some observers have argued that the Rensselaer Institute, which was founded in 1824 and began offering a degree in civil engineering in 1835, was the "most notable and influential venture in technical education" in early-nineteenth-century America. Its organizers patterned the polytechnic after similar institutions in Europe, making it distinct from established American colleges and universities. Cornell historian Morris Bishop labeled the Rensselaer Institute the "first school of civil engineering in the English-speaking world." After his graduation, Fuertes won acclaim for two civil engineering projects. He oversaw the United States government's survey of a possible isthmian canal in Nicaragua, and he consulted on the Croton waterworks in New York City.[11]

Upon arriving at Cornell in 1873, Fuertes modernized the university's civil engineering instruction, making it more than a traditional short technical training course. He revolutionized engineering education by instituting a rigorous four-year undergraduate program, which included the study of surveying, metallurgy, stone cutting, bridge construction, and geodesy. He believed that classroom participation alone was insufficient in preparing prospective civil engineers and required students to practice in the field before they left school. Accordingly, students participated in laboratory exercises and summer surveying camps. Fuertes also argued that it was the role of all colleges, especially those in the "technical fields" of engineering, to educate students as well as to train them. He required a liberal sprinkling of the arts—literature, history, and languages—in his curriculum. In a short time, he had enlarged the faculty of one to a handful of professors, including Irving Porter Church, a civil engineer and leading author of texts on engineering mechanics. It was in this setting that Anson Marston received his post-secondary education.[12]

Fuertes' program took shape during a period that was labeled "the great formative epoch in the American engineering education." The profession separated into its main branches between 1870 and 1885, and the curriculum evolved from a science course at a classical college to one that featured numerous technical aspects. The Society for the Promotion of Engineering Education believed that a major reason for the surge in engineering's popularity at this time was the transformation of America from an agricultural to an industrial nation. Following the

Civil War, activity increased in the building of railroads, metal bridges, municipal waterworks, sewerage systems, and mineral exploration and processing. Many sought employment in these fields, but few had the expertise to make noteworthy contributions because they lacked basic education in engineering. In reflecting in 1896 on the emergence of this phenomenon, Robert Fletcher, director of the Thayer School of Civil Engineering at Dartmouth College, claimed that by the 1870s, the field of engineering was widening. "New materials and forms of construction were coming into use; new experimental data were needed; hence," Fletcher argued, "the proper scope for an engineering course was rapidly enlarging."[13]

Marston, as one of Fuertes' undergraduate students, devoted his summers to practical engineering by working as a surveyor's assistant for the Illinois Central Railroad, as McCullough was to do a generation later. He graduated from Cornell with the degree of civil engineer in 1889, and went to work with the Missouri Pacific Railroad, beginning as a draftsman. For the next two and one-half years he was assigned jobs of increasing demand and responsibility, moving up to the field engineer's position. In 1891, he became construction engineer for a major drawbridge for the company over the Ouachita River, near Columbia, Louisiana. Marston left the Missouri Pacific Railroad in 1892, when he accepted an invitation to become professor of civil engineering at the Iowa Agricultural College, in Ames.[14]

Even though Iowa's land-grant college began offering a mechanical engineering curriculum to the school's first class in 1869, the courses did not resemble a field of study, except that they were heavily weighted toward mathematics. By the early 1870s, the college reflected the national trend by starting to provide specialty courses in civil engineering and in 1877 established a department of civil engineering. But the program showed little promise during the next fourteen years. Academics who directed it either possessed insufficient teaching experience or did not remain long in Ames.[15]

In the 1890s, the civil engineering curriculum improved. The state Farmers' Alliance protested that Iowa's Agricultural College was failing in its mission as a land-grant school of higher education. It charged that while engineering and veterinary science education were satisfactory, the school's efforts in agricultural instruction were dismally insufficient. Bowing to criticism, college President William I. Chamberlain resigned, paving the way for fundamental institutional changes.[16]

In 1891, William Miller Beardshear became the new president. He had proven himself as a leader of educators through his management of the West Des Moines school district. His closest supporters believed him to be a "man of experience, adaptability, and personal appeal. . . ." Beardshear, together with his recently appointed professor of agriculture, James Wilson, made sweeping changes in the college's agricultural program, and they seized the opportunity to make additional faculty appointments throughout the institution.[17]

President Beardshear's selections for two mechanical engineering vacancies and one in civil engineering were all graduates of the progressive curricula at Cornell University. In addition to Anson Marston, Beardshear hired two graduates of Cornell's renowned Silbey College of Mechanical Engineering for the two mechanical engineering posts. Beardshear's actions brought to Ames three of the brightest young engineers in the country, who helped create respectable fields of engineering study for the Iowa school.[18]

When Marston's career at the Iowa Agricultural College began in 1892, only a few enrolled in the civil engineering course, which was fortunate because Marston's only help was a half-time instructor. Shortly thereafter, student enrollment rose dramatically and the staff expanded. By 1910, there were four instructors in civil engineering. National enrollment increases do not solely account for the upsurge in engineering students at what was now Iowa State College (ISC); it was due, in part, to Marston's dynamic, progressive leadership in promoting an important field of study in an increasingly complex society.[19]

Like his mentors, Marston perceived the need for a well-rounded, thorough engineering curriculum that would instill in students the underlying principles of the field while providing a broad-based education. A school could not "make engineers of its graduates," but it could, Marston argued, "fit them to become engineers." Students could receive an education through college course work, he explained, but "many things about the details of practice can only be properly impressed on the understanding and memory [of students] by the actual performance of engineering duties."[20]

Marston believed that the curriculum in engineering schools should devote half of its time to instruction in the basics of many fields of knowledge. In mathematics, these included algebra, plane and solid geometry, analytical geometry, and calculus. In the sciences, his pupils studied chemistry, geology, and botany, and the various sub-fields of physics. Marston was also convinced that rhetoric, French or German, and cultural studies should round out the "non-technical" course work, because they broadened the students' intellectual horizons. The remaining credit hours he reserved for "technical work," such as the study of structures, laboratory experimentation, and the practical skills of fieldwork, which he saw as opportunities to apply the abstract to real-life situations. Marston wrote that he hoped to "train the student's eye and hand as well as his intellect." His formula for engineering education embodied progressive views of society.[21]

Years later, Marston reflected that, when he began his career in Iowa, the school was "practically unknown to the engineering profession of the United States, including engineering educators." He added that "the number of students was small and our equipment was scanty." Within six years, Marston had transformed the civil engineering curriculum. Nonetheless, a colleague from another institution commented that he believed Marston was "crazy" to refuse an offer from another more prestigious eastern university and remain at Iowa State.[22]

DURING HIS FIRST YEAR AND A HALF at Iowa State, McCullough mastered mechanical drawing. Sometimes referred to as "descriptive geometry" or "graphic theory," it taught students how to convey design ideas concisely with pen and paper. This included problem solving through creating axonometric projections—perspective drawings of buildings or structures from detailed plans. Closely related was surveying, which most professors recognized as a basic part of the field. Marston regarded drafting and surveying as the "A, B, C's of engineering, just as essential as a knowledge of reading and writing." Students gained practical experience with delicate instruments and the opportunity to apply their mathematical knowledge. Marston believed that drafting and surveying taught precision and provided a better feel for "the use of the tools of the profession."[23]

Marston, like Fuertes, theorized that civil engineering students needed basic experience in the field. By 1898, he had established a summer camp where students conducted a topographical survey of some region of the state, beginning on the Monday before spring commencement, and continuing for two weeks. Students paid for their own transportation and meals. The survey conducted at the camp began where the previous year's group had left off, until several square miles had been mapped.

Students usually attended two or three successive years, with upperclassmen serving as coordinators. This opportunity to use skills learned in class, Marston believed, was essential.[24]

Marston allowed students to substitute for summer camp at least four weeks of "actual engineering work done for some competent engineer, a reputable firm, or department engaged in engineering work." He encouraged students to find summer employment in their field of study because it gave them a chance to apply their academic learning and enabled them to work with experienced professionals. Because the field was advancing so rapidly, Marston believed that educators could only train students in its general principles; practical experience, as an extension of the classroom education, was, in his estimation, invaluable.

McCullough participated in Marston's summer camp program for three seasons, even though he had already worked on railroad survey crews. His previous roles as rodman, chainman, and transitman proved valuable.[25] In the summer of 1907, McCullough worked for the Illinois Central's Springfield Division at Clinton, Illinois. From September 1907 to March 1908, he found a part-time job with the Indianapolis Southern Railroad as a surveyor's helper, laying out rail lines. During the summer of 1908, he returned to the Illinois Central Railroad, but because of his standing and interest in structural engineering, he was assigned to the line's Omaha Division bridge construction operations. The following summer he broadened his skills by working for the city of Ames as construction engineer for its municipal sewer project.[26]

By the end of his junior year, McCullough was well grounded in civil engineering. Under Marston's direction, he now understood mechanics, electricity and magnetism, construction materials, and qualitative analysis. He had become proficient in differential and integral calculus and analytic geometry. He also had experience in small civil engineering construction projects. By his senior year he had to choose among railway, hydraulic, and structural engineering as a field of specialization.

Previous experience with railroads and the city of Ames directed him toward the study of structures during his final year. He mastered shear and moment analysis, and he could calculate stress in designing bridge and roof trusses. He studied the various types of metal bridges popular in the early twentieth century, including the steel arch, cantilever, suspension, and swing spans. He also became familiar with theories behind foundation analysis for these and masonry structures. His course work included road engineering fundamentals such as pavement types and maintenance costs, and he enrolled in a mandatory course on the history of engineering. Finally, McCullough became familiar with arch bridge designs constructed of stone, brick, and reinforced concrete.[27]

An essential theme in his training was that civil engineering had an economic basis because projects often relied on taxpayers' money. Designers could not risk proposing undertakings that a tax-weary public might reject. McCullough's thinking on this topic was most influenced by John Edward Kirkham, a recognized authority on arch and movable bridge design and construction. He arrived at Iowa State in 1907 as an associate professor of structural engineering. Trained in civil engineering at the University of Missouri, Kirkham had worked for twelve years in the private sector as a draftsman for a well-known bridge consultant, John Alexander Low Waddell, of Kansas City, and for the Carnegie Steel Company of Pittsburgh, Pennsylvania.[28] While at ISC, Kirkham wrote a textbook on structural engineering based on his years with some of the world's greatest designers. McCullough and his classmates learned much from Kirkham's experiences. Most important, they understood "the complexities of bridge planning and construction—both in the lab and in the field—and the overall consideration of economic principles in design."

McCullough was in the classroom at a time when concrete bridge construction was becoming a popular form for railroad and highway structures. In addition, several differing philosophies for reinforcing concrete appeared.[29] Kirkham lectured McCullough and his classmates on late-nineteenth-century European and American advances in reinforced-concrete arch construction. He showed them that solid unreinforced-concrete arches functioned well in compression, but that movement caused by live loads often created bending and buckling forces, or destructive tension, within the structures, causing them to fail. Greater mass absorbed these tensile stresses, but it was inefficient and uneconomical. Elastic metal components of wrought iron or steel used to reinforce concrete bridges absorbed tensile and shearing stresses while the rigid concrete continued to sustain compressive forces. Reinforcing helped add stability to the structures at a cost less than the additional concrete.[30]

Engineer Ernest L. Ransome improved upon a European system of wire mesh reinforcing by substituting twisted steel rods. His 1889 Alvord Lake Bridge in San Francisco's Golden Gate Park was the first American arch span with reinforcing bar. Others, like American Daniel Luten, attempted to improve upon this system. Meanwhile, Viennese engineer Josef Melan received American patents, in 1893, for a reinforcing system using steel I-beams curved to the arch axis and laid parallel, side-by-side, near the underside of the arch. The first bridge of this type in the United States was a 30-foot span built in Rock Rapids, Iowa. Melan's approach was popular with many bridge builders in the early 1900s because it was not a radical departure from the traditional concrete arch form. Nevertheless, it proved less economical than Ransome's system because of the prohibitive amounts of steel it required, often making its bridges little more than metal structures with concrete envelopes. Moreover, concrete bonded unsatisfactorily with the smooth surfaces of steel.[31]

Marston asked his senior students to write a bachelor's degree thesis. He saw this assignment as an opportunity to go beyond the classroom, where the object was instruction and recitation, what he called the "ordinary work." He intended the thesis "to demonstrate the student's ability . . . to make, on his own responsibility, a thorough study of some engineering question or problem. . . ." He was particularly interested in sharpening the skills of investigators, refining their ability to conduct experiments on some engineering question or problem.[32]

Marston directed the students to look at elementary questions, believing that they did not have enough experience as engineers to carry out in-depth investigations of complex structural problems. For the same reason, he would not allow them to design structures as their thesis exercises. He argued that "no student is an engineer when he graduates, and can only become an engineer after long years of experience." Theses involving original structural designs "were apt to produce the poorest results." Marston's goal, in addition to encouraging his students to complete independent studies, was to spark their interest in engineering research, to investigate and question the field with the hope of improving it.[33]

McCullough and a classmate, H. B. Walker, chose a simple experiment as their senior project: they researched the effects of external temperature variation on concrete bridges. They found that the range of ambient temperature and direct sunlight affected the internal temperature of bridges and caused expansion and contraction of the structures. This placed varying levels of stress on particular components, possibly causing early structural failure if not accounted for in the design.[34]

McCullough received his bachelor of science degree in civil engineering from Iowa State College in June 1910. He graduated as one of Marston's most qualified students, and his standing reflected academic distinction. By this time, McCullough had become fascinated by bridge construction, more particularly reinforced-concrete spans. Kirkham's courses on structures had caught his interest, particularly those dealing with arches and reinforced concrete. His next fifteen months as assistant engineer for Marsh Engineering of Des Moines drew him ever closer to a deep appreciation for bridge construction that was economical and aesthetically pleasing.[35]

James Barney Marsh was the owner and chief engineer the Marsh Bridge Company, which had been founded in 1904 and later reorganized as the Marsh Engineering Company in 1909. Marsh had worked as a bridge designer and marketer for the King Bridge Company, at its Des Moines office. In 1896 he formed his own consulting firm and began focusing on reinforced-concrete structures. Marsh created bridges in numerous Midwest cities, and by 1910 he had developed two varieties of reinforced-concrete through arches that he built throughout Iowa, Minnesota, and Kansas. The structural components of these bridges were a departure from the Melan and Ransome reinforcing systems, consisting of a core of steel lattice surrounded by concrete.[36]

Marsh hired McCullough in the summer of 1910 to help him design several structures for the Iowa State Highway Commission. For McCullough, it was an opportunity to learn from an expert practitioner about the economics and aesthetics of publicly financed road bridges. Marsh was busy designing and patenting what would become his signature "rainbow arch" bridges. McCullough's experience with Marsh influenced his own outlook on highway bridge construction.

Marsh's two types of reinforced-concrete through arch on first glance appeared quite similar in design. In reality they varied greatly because he created them to meet different conditions at stream crossings. The first consisted of two arch ribs rising parallel from an abutment, spanning the

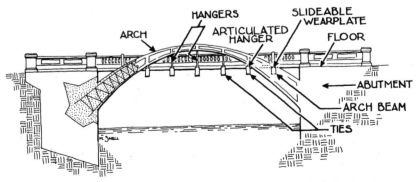

Marsh Arch. (*Kansas Preservation,* Kansas Historical Society)

stream, and falling toward the pier on the other side. Hangers suspended the floor, or road deck, from the arch ring. Slidable wear plates facilitated movement of the endmost deck beams as the arch rose and fell due to longitudinal expansion and contraction caused by changes in temperature.[37]

In the second type, Marsh created a "bowstring" or "tied arch." It differed from traditional structures because the arch ring and road deck functioned as a continuous, integrated unit. It was analogous to an archery bow and string, with the arch ring representing the bow and the road deck symbolizing the string. It was a design akin to the old metal bowstring arch bridge that the King Bridge Company had perfected in the post-Civil War period. The road deck contained the ring's horizontal thrust action, eliminating the need for heavily reinforced piers. Instead, the tied arch rested on lightly constructed piers that did not require solid rock foundations. This design was popular at sites with shallow, sandy streambeds. Marsh believed that the reinforced-concrete arch bridge was the span of the future. He claimed that all of his designs held an advantage over more traditional wooden or steel truss spans because they were resistant to floods, frosts, and fires.[38]

From June 1910 through August 1911 (when he resigned from Marsh Engineering at Marston's request to become chief draftsman for the ISHC), McCullough worked for Marsh as the latter was developing his signature spans. It was an opportunity for McCullough to continue his learning about economic highway bridge types and to apply much of what he had learned during his four years at Iowa State College. In addition, he experienced the private side of engineering, including rival companies' competition to win bridge-building contracts, and came to appreciate the arch form as one approach to building an economical and aesthetically pleasing bridge.

Conde B. McCullough continued to sharpen his skills and increase his expertise in road and bridge building, but he was beginning to understand that people thought of bridges as more than merely the means to travel from one side of a stream to the other.

❦ 3 ❧

Building Bridges for Iowa, 1911-16

Anson Marston, Dean of the Division of Engineering at Iowa State College "speaks of Mr. McCulloch [sic] as 'one of the brightest men that ever graduated from the [ISC] civil engineering department.'"

Barometer
Oregon Agricultural College
19 September 1916

WHEN CONDE B. MCCULLOUGH left the Marsh Engineering Company in 1911 to begin work with the Iowa State Highway Commission (ISHC), he embarked on a career in public service as a member of a young and elite group of college- and university-trained civil engineers scattered throughout the United States as staff members of state highway departments. On the eve of World War I, the federal Office of Public Roads and the Federal-Aid Road Act united them in a mission to build a network of highways for the American public.

From 1911 to 1916, McCullough sharpened his skills as a civil engineer, more particularly as a structural designer and researcher. He was part of a young professional staff at the ISHC that included some of the nation's experts in road and bridge building. McCullough dedicated himself to improving Iowa's poorly planned highway structure that primarily provided farmers access to markets and only secondarily served an increasingly mobile public.

The movement for better roads in the United States had started in earnest nearly twenty-five years earlier and influenced McCullough's chosen career. In the early 1890s, the League of American Wheelmen (LAW), an association of bicyclists, established the National League for Good Roads (NLGR), which promoted national highway improvements. Members believed that improved roads would end rural isolation, stimulate economic growth, and enhance education. These organizations experienced their first success in 1893, when Congress funded a federal investigation of road construction and management. The new Office of Road Inquiry (ORI), as part of the U.S. Department of Agriculture, became the federal government's participant in the country's good roads movement.[1]

The ORI's small staff became a clearinghouse for general information concerning road laws, paving materials, and technical topics that it shared with local governments, interest groups, and the public. But the ORI's most valuable function was the "object-lesson construction program," in which federal engineers built short experimental sections of smooth hard-surfaced roads. Engineers demonstrated the newest construction techniques to the public. The program spread the gospel of good roads to all who would listen. Working with the LAW and the NLGR, the ORI intended to educate and persuade the public so that it would support improved highway networks throughout the country. The creation of Rural Free Delivery mail service in 1896 convinced farmers who had once sneered at a major overhaul of the nation's road system. Farmers became vocal proponents because they benefited from both programs.[2]

In 1899, the ORI was designated the Office of Public Road Inquiry (OPRI) and with energetic leadership sought greater publicity for the advantages of improved roads. The object-lesson road demonstration program expanded. In addition, the agency distributed hundreds of thousand of copies of circulars promoting the benefits of state monetary aid to county road-building programs. According to Bruce Seely, by 1904, all states east of the Mississippi River except Indiana provided some assistance for such activities. New England states offered substantial support, as did Washington state, which paid 50 percent of state road construction costs. But most others, including Iowa, gave little money and even less incentive. Nonetheless, the "Good Roads Trains," operated by the OPRI and sponsored by rail companies who saw the potential for greater profits through better field-to-siding roads, promoted highway improvements, and cultivated the strong farm lobby for a system of well-maintained rural routes.[3]

In 1904, the local good roads association finally overcame opposition in the Iowa General Assembly, which passed a law that created a state highway commission. The association, including one of its strongest defenders, Anson Marston, was most concerned about farmers' difficulty delivering their produce to market and characterized Iowa's administration of its 100,000 miles of roads as poor. Similar state groups lobbied lawmakers all across the country. By 1904, they had helped create sixteen highway commissions. The motive in Iowa and elsewhere was to assist farmers, not to provide a system of roadways for automobiles, for these still smacked of elitism. The bicycle and the automobile were, at best, allies in the drive for improved market or post roads and a nationwide highway system.[4]

The ISHC created in 1904 was, for all practical purposes, Iowa State College and its administrators. Public sentiment favored improving state roads, but not the establishment of a separate state highway department with its own jurisdiction. Under the guidelines of the 1904 law, the school's board of trustees appointed Anson Marston and C. F. Curtiss, dean of agriculture, to carry out the commission's work. The inclusion of Curtiss fit well with the mission of constructing properly maintained roads for transporting farm products to markets. Neither professor was given the authority to conduct or oversee construction activities; the commission was to simply "devise and adopt" suitable methods of highway construction and maintenance, to provide county road officers with information on how local governments might improve travel.[5]

In the summer of 1904, Marston canvassed several Atlantic coast and Southern states seeking an experienced engineer to head the commission's activities in Ames. Finding no one suitable, Marston convinced the state to hire one of his recent graduates, Thomas Harris MacDonald, to undertake Iowa's road investigations.[6]

MacDonald had previously worked summers in various capacities for railway companies, and for Marston in drafting plans for sewage disposal systems. His investigation, conducted on horseback, sparked significant interest in the local good roads movement and his research made him one of the nation's experts on the current theories of roadway construction and management. Marston urged MacDonald to continue his investigation of Iowa's road system beyond his bachelor's degree thesis. As the commission's staff member, his goal was to recommend a solution to Iowa's muddy dilemma.[7]

Thus, Thomas MacDonald became the ISHC's only paid employee, at an annual half-time salary of six hundred dollars. He continued his Iowa road survey limited by a small budget, answering occasional inquires from the few county road supervisors who sought his advice and suggesting improvements to county supervisors and provided standard plans, but not much more. An observer noted that the commission consisted of "one man, a desk and two filing cabinets, all occupying less than 90 square feet of floor space in a corner of a room on the fourth floor of Engineering Hall at Iowa State College." MacDonald's budgetary predicament paralleled what was happening in other states whose highway commissions had no state funds for road building projects.[8]

Meanwhile, beginning in 1905, the Office of Public Roads (OPR), the OPRI's successor, increasingly promoted its information-gathering activities, under the leadership of Logan W. Page. The new director vigorously sought cooperation between state highway commissions and the OPR, seeing networking as crucial to disseminating knowledge and promoting high-quality roads. Page believed in the Progressive philosophy of using a disinterested apolitical expert;he thought that by avoiding the politician he might make expensive road construction more palatable to the American public. MacDonald, Marston, and Curtiss hoped to capitalize on Logan Page's conviction in completing Iowa's road study. Their most urgent task was to assess the condition of the state's bridges and culverts, old and new, because it was uncertain that these structures could carry even their own weight, or dead loads, let alone traffic.[9]

By 1907, the lukewarm response to highway commission activities had passed and local officials and the public embraced state-assisted road construction. The day of the engineering specialist as a leader in promoting good roads in Iowa had arrived. MacDonald learned that counties spent all the tax revenues allotted to them for bridges, mostly constructed of substandard timber or light steel. He and his office sponsored "road schools," which informed county supervisors about the advantages of grading methods to prevent erosion and to promote drainage of perennially muddy stretches of county roads. MacDonald's highway commission duties became so time consuming that he began to work full-time as the ISHC's highway engineer. In his expanded role with the commission, he maintained close ties with the state college's civil engineering department through the college's engineering experiment station. Marston and the General Assembly had begun the experiment station three years earlier to conduct scientific investigations, test and analyze materials, and produce technical information on engineering topics. The ISC experiment station staff cooperatively researched several durable paving materials for Iowa's roads.[10]

Several regional construction companies held monopolies on bridge-building contracts in some Iowa counties; company agents often persuaded county officials to accept noncompetitive bids for construction of all spans built during a fiscal year. The ISHC received frequent complaints about these structures because of their poor quality, light load capabilities, and high costs. But because the highway commission functioned in only an advisory capacity, there was little it could do. It could not prevent counties from granting blanket contracts to certain companies, nor could it require the use of standardized plans for bridge and culvert construction or plans the commission had reviewed and approved.[11]

Nonetheless, by 1911, local demand for standardized plans for culverts and bridges had increased to such an extent that the commission was forced to employ more personnel. It hired the twenty-four-year-old McCullough as a draftsman. Since graduating in the spring of 1910, he had showed skill and ambition as a bridge designer with the Marsh Engineering Company of Des Moines. In Ames with the ISHC, McCullough created a collection of standardized plans for spans that provided safe stream crossings for rural Iowa roads. He hoped to satisfy a backlog of requests from county government. McCullough was creating the structures, not just sketching someone else's ideas. He was promoted to "designing engineer" of bridges for the highway commission and accordingly he "became the first assistant engineer to Mr. MacDonald, with a regular desk and title." That same year, the commission outgrew its ninety square feet of office space and moved to larger quarters on the ground floor of Engineering Hall. Funding was still meager, but the full-time staff now totaled three, with the addition of an office assistant.[12]

As designing engineer, McCullough prepared a large collection of drawings and specifications that would serve as the highway commission's standardized models for small bridges and culverts. These ranged from relatively simple box girder spans to larger reinforced-concrete arches. County engineers received the plans free of charge and used them for "in-house" construction or work contracted to outside builders. The ISHC hoped the program would improve the overall quality of Iowa's bridges and culverts.[13]

The Iowa General Assembly gradually increased the ISHC's annual appropriations so that by the early 1910s they amounted to $5,000. Marston termed the legislators "unprogressive" because they remained blind to the need for a better-funded highway commission. One county saved enough "on the construction of bridges," Marston noted, "to pay

Engineering Hall, Iowa State College, 1918. (Iowa State College, 1918.
Iowa State University Library/University Archives)

the entire amount of maintaining the expense of the commission from
the time it was first established."[14]

In 1911, Fred R. White and John H. Ames, new ISHC employees and
both ISC graduates, began a comprehensive evaluation of the economic
efficiency of county governments' role in road and bridge construction,
conducted under McCullough's guidance. They found most projects to
be a waste of taxpayers' money. In one county, the board of supervisors,
together with interested citizens, embarked on a project to grade and
gravel nine miles of road located on low-lying land with poor drainage.
Plans called for hauling gravel to the site, but ignored subsurface
preparation, resulting in a road that very shortly became a muddy,
impassable trail.

A far more flagrant waste of tax revenues was the blanket contract
system of bridge construction; counties that disregarded the ISHC's
request to adopt standardized designs continued to award bridge work
to single building firms.[15] The general design and location of each
structure was left to the company foreman's discretion. Detailed plans
for county officials' inspection were generally nonexistent; those that
did exist were often incomplete. Regional builder Daniel B. Luten used
his own patented designs, but Thomas MacDonald reported that,
"Probably ninety percent . . . were privately let without other plans or
competition being considered." Bridge construction costs spiraled, with
addition after addition made to the initial estimate. In many cases, there
was no written contract, and expenditures exceeded anyone's expectations
except the company's representative. County road officers were "almost

Substandard bridge in Iowa, 1914.
(Iowa State Highway Commission,
Courtesy of Iowa State University Library/University Archives)

Substandard bridge in Iowa. Built 1910, collapsed 1912.
(Iowa State Highway Commission *Service Bulletin*, 1914.
Courtesy of Iowa State University Library/University Archives)

Bridge unsuitable for site, 1912. Wingwall cost as much as rest of bridge.
(Iowa State Highway Commission. Courtesy of Iowa State University
Library/University Archives)

universally absolutely honest and [had] the best of intentions," Marston believed, but the system was flawed because county officials lacked professional training as engineers and thus could not make informed construction decisions.[16]

Compounding the situation, some costly bridges were ill suited for the locations on which they were built. White and Ames found that some structures were too small to provide sufficient clearance for high water under the road decks. Strong currents frequently widened river channels and cut around bridge abutments, rendering the spans useless. In the other extreme, some bridges with heavy abutments, retaining walls, and high riveted steel spans cost more than $5,000 when $500 reinforced-concrete culverts would have accommodated both road traffic and stream flow. Bridge company foremen repeatedly erected spans at entirely inappropriate locations. White and Ames cited the case of a forty-foot steel structure in which the bridge and three of its four wing walls cost $2,900. To make the span fit the site, a fourth wall was extended sixty feet at an additional cost of $2,200.[17]

White and Ames cited repeated instances of double charges for goods and services, and they uncovered a case of two counties each unknowingly paying the entire costs of constructing the same bridge straddling a stream that served as a common boundary. They also found "a very inferior arch bridge [being] built from carefully prepared plans for a flat top bridge." They blamed these incidents on untrained bridge construction supervisors and contractors' foremen who were "not working for the county." Iowa demanded good roads, but it often received expensive and inadequate bridges. Limited tax revenues, earmarked for general road construction, were often diverted for expensive bridges, leaving nothing for other important projects. Thus, the state had several new bridges— and unimproved roads which were nearly always impassable. McCullough's bridge study demonstrated that taxpayers needed to overhaul the state's highway program. This philosophy harkened back to Logan Page's belief that apolitical experts needed to promote the idea of good roads.[18]

Many called for reform. The weekly journal *Engineering News*, in June 1912, echoed the two Iowa engineers' views, reporting the problem widespread throughout the country, but especially predominant in the Middle West,

> *Anyone who is at all acquainted with conditions in country highway bridge building today . . . knows that we still have with us the ancient evil of selling small highway bridges like yards of calico, with no guarantee, with no technical supervision and with no authoritative intermediary between ignorant and often corrupt local officials and the complacent and conspiring bridge companies.*

Engineering News called for state supervision of all highway bridge construction, a solution it had advocated twenty years before. It sought state departments with wide authority to ensure proper bridge design and construction, to protect communities from the "venality and ignorance of [their] elected officials." State supervision, *Engineering News* argued, would not kill "honest private competition."[19]

Iowa's legislators had introduced many road and bridge bills in the General Assembly after 1904, but companies lobbied successfully against them. Revelations in 1913 about county officials engaged in dishonest practices with bridge companies persuaded the Assembly to approve legislation creating a more powerful state highway commission. The reorganization established an independent state agency, with a new commission composed of three appointees, one from each of the two major parties and a nonpartisan expert. Initially, the third member was Anson Marston.[20]

The 1913 law gave the new ISHC jurisdiction over all state roads. Counties were to select between 10 and 15 percent of their rural mileage for inclusion in a county highway system, which would become the focus for ISHC activities. The legislation authorized the highway commission to hire professionally trained staff and to dismiss unqualified county road personnel. The legislation left the commission to implement a comprehensive plan for road and bridge construction in Iowa, but failed to authorize state funding for the actual work. It was unusual in the 1910s for states to demand control without providing funding for construction costs. "Every state that is making a decided advance in the building of permanent roads," MacDonald observed, "is accomplishing this object through . . . some form of state aid." Meanwhile, local county officials incorporated ISHC direction through county-financed tax initiatives.[21]

With the ISHC reorganization, MacDonald continued as state highway engineer and White became "field engineer," a post in which he supervised five district state engineers. They worked closely with road officers in twenty of Iowa's ninety-nine counties in solving construction problems. The commission also employed an "educational engineer" who assembled staff members' publications, which addressed practical questions about road-building techniques. The commission directed these circulars and bulletins to county officials. McCullough became the assistant state highway engineer, in charge of the design department. With a staff of eight to twelve draftsmen, including John H. Ames, he prepared general plans for bridges and culverts. McCullough also engaged in pioneering

Large bridge designed by Iowa State Highway Commission using standard plans. (Iowa State Highway Commission. Courtesy of Iowa State University Library/University Archives)

research, including a large project in which he examined bridge placement in relation to topographic and climatic concerns.[22]

In their first research undertaking, twenty-five-year-old McCullough and his assistants studied the topography of the state to gain a better understanding of the action of glacial sheets in the formation of its predominant flat plain and rugged northeastern sections. This was essential because it determined the selection of bridges for each project. McCullough concluded that the Iowa plains comprised three distinct regions. The Kansas drift in the southern half of the state had created deep ravines and the retreating glacier had removed loose, gravelly debris centuries ago. To the north, the Iowa drift, somewhat younger, left shallow streambeds. The northeast portion of the state lay in the Wisconsin drift, typified by marshes, swamps, gumbo mud, and shallow stream channels.[23]

For each zone, McCullough tailored the deck girder bridge, regardless of the length of a proposed span, to account for the regional variation in soil conditions and streambed formations. In the Kansas plain, for example, because of the scarcity of gravel for use in concrete, it was necessary to haul this material from a source in the Wisconsin drift region, over poor roads and at great expense. For the deeply cut riverbeds in this region, McCullough designed deck girder bridges to span the ravines that were relatively light in their concrete content, but heavily reinforced with steel rods. This design kept costs to a minimum, but gave taxpayers a sturdy, serviceable product.[24]

Conversely, in the Wisconsin drift, where streambeds were shallow, McCullough used fill at the banks of the rivers and creeks to create spans with enough headroom to accommodate the stream flow, even during freshets. Aggregate, or gravel, for concrete and fill was plentiful, minimizing costs for material.[25]

Finally, different types of soils found in the three drifts regions presented a variety of erosion considerations. Younger streams have a greater propensity for cutting deep streambeds, so he took precautions to avoid having foundations, piers, or abutment walls undermined by the flow of water. McCullough believed his approach would prevent the horrendous problems found with the bridges that had been erected without regard to topographic and geologic considerations. His overriding concern was that his structures be functional, aesthetically pleasing, and economical.[26]

McCullough had an affinity for the arch form and promoted it as a design type "with a true regard for safety and economy." He believed that an understanding of stream behavior was essential for a designer to create a permanent structure. He argued that design costs were far less expensive than building a bridge, though of worthy design, that was unsuitable for a particular location. McCullough's previous two years of pioneering work on stream behavior gave him a thorough understanding of the subject.[27]

McCullough, as head of the designing department, now renamed the bridge department, needed an assistant engineer and he. hired Earl Foster Kelley. An Iowa native, Kelley had earned a bachelor of science degree in civil engineering from ISC and subsequently worked for the American Bridge Company designing steel railway and highway bridges.[28]

Increasingly, McCullough channeled his research to projects involving his personal interests and expertise—reinforced-concrete bridges. He delegated the design of the steel truss spans to Kelley. Both men worked diligently to produce workable drawings of proposed bridges in an orderly fashion. They carefully recorded expenses for each job so that the commission could justify the cost-saving features of centralized road and bridge design for the state of Iowa.[29]

Another project for the two men during this period was a return to the research project that McCullough had helped initiate in 1909 as an undergraduate, when he and H. B. Walker had collaborated on their bachelor's degree thesis.[30] McCullough and Walker had hypothesized that temperature fluctuation had an effect on the stress exerted upon the entire span of the typical reinforced-concrete arch bridge. They suggested that, over time, contraction and expansion caused by these temperature fluctuations would fatigue metal reinforcing bars or plates and significantly reduce the bridge's load-carrying capacity. They charted daily variations of the internal temperature of arches comparing the influences of wind, sunlight, shade, and precipitation over the course of a year and concluded that engineers who designed reinforced-concrete arch bridges needed to account for environmental effects. Walker and McCullough's work, though of an elementary nature, became the basis for an Iowa Engineering Experiment Station *Bulletin* a few years later.[31]

Since the days of his preliminary research with Walker, McCullough had matured as a researcher and saw the need to re-examine the perplexing problem of structural temperature variation. He hoped, this time, to systematically apply his research to the design of economic bridges. He and Charles Sabin Nichols, assistant director of the Iowa Engineering Experiment Station, a colleague and fellow alumnus, again considered the effects of temperature on reinforced-concrete arch bridges. They intended to establish minimum standards for construction of this type of bridge on Iowa roads.[32]

They first canvassed civil engineers nationwide for parallel research. They learned that considerable data existed, but that "very little was known concerning the actual internal variation of temperature in concrete structures." McCullough and Nichols believed that most designers simply overbuilt bridges to compensate for the structural instability that temperature variation could induce. This was wasteful; engineers could create strong, stable, and economic bridges.[33]

The bridges had to be strong enough to carry both dead and live loads, that is, their own weight at given points and that of traffic passing over them. Designers had to understand several factors when creating these spans. For instance, the range of ambient temperature at a given latitude throughout the year, or even within a twenty-four hour period, causes bridge components to expand or contract. inducing bending stresses within the arch rings that most engineers found difficult to determine. To assure structural safety, engineers routinely overestimated the immeasurable variable, compensating for it with much additional reinforcing and concrete. McCullough argued that engineers faced this situation because "too little information . . . [was] available concerning the action of the arch type [of bridge]. . . . Very few arches are economically designed," he continued, "and the great need of experimentation along this line is apparent to all who are familiar with the problems presenting themselves to the designer." The journal *Engineering and Contracting* praised McCullough's pioneering work in this field and featured a discussion of his research in a November 1913 issue.[34]

Because of his reputation as a respected structural engineer, McCullough's role with the ISHC in studying economic bridge types continued to expand. In 1914, the commission encouraged further cooperative research with the Iowa Engineering Experiment Station staff at the State College and sponsored a general study of the safety and costs of reinforced-concrete arch bridges built in Iowa prior to the comprehensive bridge standards program.

During his years with the ISHC, McCullough investigated almost every type of bridge used on Iowa's highways. He considered the advantages and disadvantages of each according to its construction costs, maintenance expenses, and the topographic considerations of each site. The reinforced-concrete arch was his chosen design medium: bridge types were incomplete, in his opinion, unless they included the reinforced-concrete fixed arch. John Kirkham, his mentor, and James B. Marsh, his employer, influenced McCullough's preferences. McCullough embraced the arch form from an architectural standpoint "because of the quiet, simple dignity of its lines." On the other hand, he did not reject other bridge types if they met his high standards for durability and economy, and if they fulfilled his criteria for type selection with respect to site. McCullough was a committed research engineer, and his examination of internal temperature changes of reinforced-concrete arch spans was the beginning of important scholarship on economic bridges.[35]

One of McCullough's greatest challenge as a bridge engineer in Iowa came in 1914, when his superiors asked him to gather evidence in support of litigation involving a patent infringement case. The case had originated in 1900 when Daniel B. Luten, founder and president of the Luten Engineering Company and the National Bridge Company, began patenting many components of reinforced-concrete arch bridges. Over the next fifteen years, Luten applied for at least seventy-five different patents, and the U.S. Patent Office granted forty-five for "improvements" in the method of building reinforced-concrete bridges. These included over four hundred specific claims. Luten customarily received a 10 percent royalty fee from bridge contracts that employed his patented design components. He once quipped that "it is a serious question whether a reinforced concrete arch . . . [could] be erected without infringement" on his claims. Luten devoted enormous amounts of time to litigation in the midwestern and eastern U.S. trying to collect royalties on bridge contracts.[36]

Luten filed suit in federal court in 1912 against the Marsh Engineering Company of Des Moines, Iowa, alleging that James Marsh had illegally used Luten's designs in a reinforced-concrete arch bridge at Albert Lea, Minnesota, about one hundred miles from Ames. Iowa's governor directed the state's attorney general to appear in defense of Marsh Engineering in the case of *Luten v. Marsh Engineering Company* because a 1913 statute required the state to defend Iowa bridge contractors who were the objects of patent suits. The attorney general employed Wallace R. Lane, a patent attorney from Chicago, Illinois, to investigate Luten's claims. Thomas MacDonald assigned his bridge design expert, Conde B. McCullough, to assist Lane.[37]

The ISHC read with great care Luten's charges against Marsh, a trusted professional who had worked with the commission as a consultant and whom the staff engineers respected as a first-rate bridge designer. They concluded that Luten's assertions against Marsh were completely unjustified and that Luten interpreted his patent letters too broadly and sought royalties for items that were in the public domain. It maintained that Luten was making a royalties grab at taxpayers' expense.

In addition, McCullough, who was thoroughly familiar with the type of bridges that Luten designed and constructed, concluded that without heavier sections of concrete and additional reinforcing steel, they did not even meet the ISHC's standards.[38] Luten dismissed the commission's findings, and in preparing his case he obtained the independent appraisal of a University of Wisconsin research engineer to confirm his belief that his own designs were of superior quality. The professor, however, agreed with McCullough. Luten then made defamatory statements against the ISHC, charging it with prejudice and unfairness in setting its standards..[39]

McCullough and his staff devoted part of the next three years to the case, one of eleven similar pending suits nationwide involving Luten. With the assistance of legal counsel, he chronicled the international development of reinforced bridge construction prior to 1900, when Luten had filed his first patent, hoping to cast doubt on Luten's claims to originality of design. The work of McCullough and his staff included six hundred pages of printed testimony and one hundred fifty exhibits, of which fifteen were bridge models. The state's goal was to crush any bridge company's attempts to unfairly and unethically collect patent royalties on taxpayer-financed bridge construction. The case was a milestone in American civil engineering.[40]

In January 1918, Judge Martin J. Wade, of the U.S. District Court for southern Iowa, ruled in favor of the Marsh Engineering Company, dismissing all eight of Luten's claims. He concluded that Luten's patents "did not disclose new knowledge, but rather mechanical or engineering details of the application of knowledge that is old." The claims did not disclose any invention and "were therefore not patentable." The court declared virtually all of Luten's patents invalid, and censured him for his unethical business practices. *Luten v. Marsh Engineering Company* led to similar verdicts involving bridge engineering patents in Kansas, Oklahoma, and Nebraska.[41]

McCullough's background and experience had enabled him to make professional judgements concerning the authenticity of Luten's claims. His brief experience with the Luten litigation, however, was not to be his last encounter with the law.

By 1916, as part of a slight reorganization of the ISHC, McCullough was promoted to the post of assistant state highway engineer, leaving bridge designing to John H. Ames and E. Foster Kelley. He then divided his time equally between experimental research and bridge patent litigation. But in the three years after the Iowa General Assembly broadened the role of the ISHC, McCullough's design department prepared drawings for more than seventeen hundred locations, totaling over $4.4 million in estimated construction costs.[42]

Also in 1916, McCullough received the degree of "civil engineer" from Iowa State College, the equivalent of a professional license in the field , with requirements exceeding those for a master of science degree in civil engineering. McCullough earned his degree by completing five years as an engineer in a professional position, and writing a thesis. He had shown his competence, and his colleagues acknowledged his published articles in which he described his theories, experiments, and conclusions on bridge design.[43]

⊁⊰

CONDE B. MCCULLOUGH LEFT the Iowa State Highway Commission in late summer 1916, furnishing no clear reasons for his departure. In five short productive years, he had accomplished significant research and design work and had advanced the quality of Iowa's road system, but he realized that he could use his talents in other locales in the United States.

McCullough moved to Oregon with his wife, the former Marie Roddan, whom he had married in 1913, and his son John Roddan McCullough. He became assistant professor of civil engineering at the Oregon Agricultural College in Corvallis, beginning his college teaching career as the sole member of the school's structural engineering program within the civil engineering department. McCullough saw the western United States as a region where, except for a few scattered places, road and bridge building were in their infancy. He moved there because he had a creative style that demanded a blank canvass where he could use his genius to its greatest potential.[44]

After instituting a comprehensive road- and bridge-building program in Iowa, Thomas H. MacDonald became President Woodrow Wilson's choice in 1919 to succeed Logan Page as director of the Bureau of Public Roads. His promotion marked a new beginning for highway improvement in the United States. He served as chief of the BPR for the next thirty-four years, directing a comprehensive national highway system sponsored with federal aid.[45]

Fred R. White, another of McCullough's colleagues, succeeded MacDonald and served Iowa as highway engineer for several years. He continued and completed the many programs MacDonald had initiated in the 1910s and subsequently cooperated with MacDonald and the

BPR in federally assisted road- and bridge-building programs. Earl Foster Kelley continued with the ISHC for four years after McCullough's departure. In 1920, he followed MacDonald to Washington, D.C., where he joined the BPR's engineering staff and became chief of the Bureau's division of tests in 1925. McCullough, MacDonald, White, and Kelley remained close professional and personal friends.[46]

In the next thirty years, Conde B. McCullough established his own national reputation as a master designer and builder who saw bridges as more than merely the physical means for crossing from one side of a stream to the other. He lived up to the acclaim that he was "'one of the brightest men that ever graduated from the [ISC] civil engineering department,'" and one of the most gifted young engineers in the country.[47]

❧ 4 ❧

Teaching at the Oregon Agricultural College and Early Bridge Building with the Oregon State Highway Department, 1916-24

The railroads usually measure a stream, and then send out a hand-me-down blueprint for a bridge to be built to predetermined standards. In Oregon our engineers have been trained to go to the stream, build a bridge for utility and economy, and at the same time design it so it will blend with the terrain.

Conde B. McCullough
8 October 1935

C ONDE B. McCULLOUGH TRAVELED from Ames, Iowa, to Oregon's Willamette Valley in mid-1916 by rail. Cross-country journeys by automobile were still almost an impossibility except for the hardiest of souls and machines. He taught for three years as an assistant professor of civil engineering at Oregon Agricultural College (OAC, now Oregon State University) in Corvallis. His predecessor at the Corvallis school, Rex E. Edgecomb, left for a similar professorship at the University of Pennsylvania. Edgecomb had graduated from Iowa State College (ISC) in 1911, one year after McCullough, with a bachelor of science degree in civil engineering. While at OAC, he focused on bridge building and completed a thesis for a civil engineering degree at ISC. Both Edgecomb and McCullough received their diplomas for the professional degree during ISC's June 1916 commencement. Because of an uncertain future for OAC's civil engineering program, Edgecomb looked elsewhere to continue his career.[1]

Anson Marston, dean of engineering at ISC, and William Jasper Kerr, president of OAC, were well acquainted with one another. Academics knew Marston for his progressive engineering curriculum and Kerr as a no-nonsense land-grant college leader who had brought high academic standards to the Oregon school since becoming its top administrator in 1907.[2]

McCullough's contributions to the ISHC's road- and bridge-building program made him a sought-after mentor for undergraduate students enrolled in an applied engineering curriculum. The federal Office of Public Roads (OPR), which directed the United States government's participation in the good roads movement, praised the ISHC's bridge standardization program that McCullough had created, believing that, because of it, Iowa possessed "the best bridges and culverts of any state in the Union." Word reached Oregon, where John H. Lewis, the state engineer, praised Iowa for its progressive 1913 road legislation that, in part, had made McCullough's bridge program possible. McCullough's successes in Iowa had preceded the young engineer to Oregon.[3]

At the beginning of McCullough's first term at OAC, the School of Engineering and Mechanic Arts abandoned its bachelor of science degree in civil engineering. In February 1914, Oregon's Board of Higher Curricula eliminated this track and other programs, effective in 1916, to avoid duplicating degree courses at the University of Oregon. In addition to saving money, the board and President Kerr were also trying to focus the land-grant college's curriculum on technical applications in agriculture and mechanic arts. The board also eliminated OAC's degree programs in the liberal arts, fine arts, and professions, fields in which the state university had already shown itself qualified, and transferred the University of Oregon's electrical and chemical engineering departments to OAC.[4]

When McCullough began teaching at OAC as an assistant professor of civil engineering, he instructed only juniors and seniors who had entered the program before the February 1914 cut-off point. Freshmen and sophomores interested in civil engineering enrolled in course work leading to bachelor of science degrees in highway engineering or irrigation engineering, the two alternatives offered in place of the more general B.S. in civil engineering.[5]

Before McCullough joined the faculty, the Department of Civil Engineering's teaching staff had numbered four. Gordon V. Skelton was its top-ranked member as a full professor. Rex E. Edgecomb was assistant professor. Samuel Michael Patrick Dolan and Dexter Ralph Smith were instructors. The 1913–14 enrollment in civil engineering was sixty-three, in a total student body of just under one thousand. After the program was canceled, only forty upperclassmen remained in the general civil engineering program in 1915–16, but the number of students in highway engineering grew. McCullough replaced Edgecomb, teaching senior courses to students in the civil program specializing in structural engineering. He also taught upper-division courses to highway engineering undergraduates, focusing on bridge types, reinforced concrete, and contracts and specifications.[6]

Progressive era faith in specialized fields of study and the nationwide romance with good roads influenced the Board of Higher Curricula's decision to promote highway engineering over a more general course. But Grant Covell, dean of the OAC School of Engineering and Mechanic Arts, believed the board's ruling was unfair: it forced students seeking degrees in civil engineering to out-of-state schools because no other Oregon institution offered the program.[7]

Under faculty and student pressure, the Board revised its position in 1917, reinstating the Department of Civil Engineering and eliminating the independent four-year highway and irrigation engineering degrees. Although the new program again offered a more general bachelor's degree, the department now gave instruction in specialty fields of structural, highway, and irrigation engineering. Skelton and McCullough, together with Thomas A. H. Teeter, professor of irrigation engineering and hydraulics, formed an executive committee that directed the revived program. McCullough oversaw the structural specialty as professor of civil engineering, Skelton directed highway engineering and Teeter headed irrigation engineering.[8]

Dolan and Smith became two of McCullough's closest life-long colleagues and friends. McCullough's cynical humor and energetic style matched the brashness of Sam Dolan who, according to one of his students, was "a big husky Irishman" who had played football with Knute Rockne at Notre Dame before joining the faculty at OAC. Dex Smith was a "feisty little fellow, very intelligent, very smart." With reorganization, Dolan became assistant professor, teaching sophomore and junior year courses on road construction and highway engineering. Smith remained the mechanical drawing instructor.[9]

⊱⊰

McCULLOUGH UNDERSTOOD OREGON'S DRIVE to "lift its feet out of the winter's mud and summer's dust," according to one Oregon State Highway Department (OSHD) chronicler. Oregonians had approved an initiative authorizing the state to finance a permanent highway system through bond issues, but they split on further road laws. Some favored a State Grange proposal to create a modest state highway department that, not unlike Iowa's first highway commission, could only advise county courts on road and bridge construction and maintenance programs. Other Oregonians supported a bill to create a more powerful state road board, consisting of the governor, the secretary of state, and the treasurer, which would appoint a commissioner to allocate a $1 million bond-financed budget for county projects. The controversy over local versus state control of road building was a familiar story replayed across the United States. At the national level, many supported creating a strong highway commission, but critics argued for a strong federal road- and bridge-building plan, with local control over intrastate matters.[10]

The OSHD's first staff, beginning in 1913, had included Henry L. Bowlby as state highway engineer, Samuel C. Lancaster as assistant state highway engineer, and Charles H. Purcell as state bridge engineer. While Lancaster was designing the Columbia River Highway, Bowlby and Purcell canvassed the state gathering information to establish a designated state road system. Along the way, they found deplorable bridges at many locations and concluded that "customary bridge [building] methods" needlessly cost Oregon's taxpayers thousand of dollars annually. Purcell reached conclusions similar to McCullough's from a few years earlier in Iowa. He noted that bridge company representatives sold county courts, which oversaw road construction and maintenance in the pre-automobile era, costly and often inferior structures. Many were ill-suited for their locations because the "courts 'fall for' the talk put up to them by these salesmen." Bowlby and Purcell sought additional legislation authorizing the highway commission to design and supervise bridge construction on all publicly owned routes in Oregon. They hoped that quality bridge construction on the Columbia River Highway, the Pacific Highway, and other state roads might garner public support for a stronger highway commission.[11]

Increasingly, Oregonians were coming to believe that a powerful highway commission could implement a comprehensive road and bridge construction and maintenance program and produce additional funding through bond issues. Oregon voters approved such a plan in 1917. But in the end, federal law provided the impetus for change.[12]

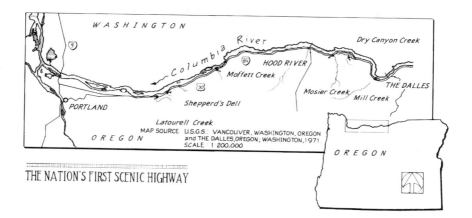

THE NATION'S FIRST SCENIC HIGHWAY

The idea of building a scenic highway along the south bank of the Columbia River was conceived by Samuel Hill and other prominent Portland businessmen. Hill took Samual Lancaster and future State Highway Engineer Henry Bowlby to Paris and the Rhine River Valley of Germany to analyze highway development in Europe. With the establishment of the Oregon State Highway Department in 1913, Sam Lancaster was hired to oversee all preliminary engineering proposals and designs prior to pavement construction. State Bridge Engineers C. H. Purcell, K. P. Billner, and L. W. Metzger created the innovative bridge designs that were constructed between Troutdale and Eagle Creek. The grade of the highway is never more than 5 percent and the average road width is twenty feet. Topographical and scenic values of the gorge were important factors in determining the bridge designs and locations. The Columbia River Highway was the first scenic highway constructed in the United States.

Latourell Bridge was the lightest reinforced concrete three-span deck arch on the highway. Shepperd's Dell is a deck arch design with a unique solid curtain wall above the center of the arch. Moffett Creek Bridge was the longest single span three-hinged deck arch in the United States in 1915. These three bridges are representative examples of the creative genius of Bridge Engineers Purcell, Billner, and Metzger.

The State of Oregon established a three-man State Highway Commission in 1917 and committed a sizeable amount of funds for the completion of the Columbia River Highway. In 1919, the Commission hired Conde B. McCullough as State Bridge Engineer. McCullough designed three bridges that connected the highway to The Dalles. The designs of Mosier Creek and Dry Canyon bridges were influenced by the rib arch form of K. P. Billner's Shepperd's Dell Bridge. McCullough assumed the work of Sam Lancaster and designed these bridges so that they complemented the scenic beauty along the Columbia River Gorge.

Columbia River Highway. (Oregon Historic Bridges Recording Project, Historic American Engineering Record [HAER], National Park Service, Todd A. Croteau *et al.*, delineators, 1990)

In 1916, Congress passed the Federal-Aid Road Act, committing the federal government, through the Office of Public Roads (OPR), to nationwide highway building. The Act appropriated $75 million in federal matching funds over several years. States would receive portions of this on a project-by-project basis for grading and paving road surfaces, and constructing bridges on post roads. But the Act required that the OPR approve state highway commissions and departments before they could participate in the program. The OPR was looking for commissions with the necessary financial resources, legal authority, and skilled employees to initiate federal-aid projects. Framers of the federal act hoped to stop county governments from constructing roads and bridges with insufficient supervision by centralizing and systematizing state highway programs. In 1916, only California's program conformed to the OPR's guidelines. Slightly modifying current laws was enough to bring some states into compliance. A few, including Oregon, completely reorganized their commissions. The OPR required eight additional states to form them from scratch.[13]

In February 1917, the Oregon legislature placed all state highway system construction and maintenance in the hands of a three-member commission appointed by the governor. It also placed on the ballot a $6 million bonding act to finance the OSHC and its department's increased responsibilities. A majority of Oregonians, disgusted with previous road laws and paltry funding, approved the referendums on a June ballot. Governor James Withycombe appointed a new board of commissioners, choosing Simon Benson, a Portland lumberman, good roads enthusiast, and early promoter and financier of the Columbia River Highway, as its chairman.[14]

The new Oregon State Highway Commission selected Herbert Nunn as its highway engineer to oversee day-to-day departmental activities. Because of wartime emergency conditions, Nunn directed a disproportionate amount of road resources to completing and improving the state's trunk routes for motor trucks transporting war materiel. These routes included the Columbia River Highway, from Astoria east to Umatilla, and the Pacific Highway, from Portland south to the California state line. Nunn also enlisted his engineers, at state expense, to assist in county road construction. He offered the bridge department's services in designing new spans free of charge to county courts, thereby ensuring that new structures met state minimum safety standards for load, width, and overall design.[15]

Charles Purcell resigned as state bridge engineer in 1915, to pursue consulting. Over the next few years, the his former post remained vacant, with staff engineers overseeing bridge construction. Meanwhile, bond sale proceeds and early federal-aid matching funds increased the pace of highway projects in Oregon, and a boost in automobile license revenues eased worries about retiring the mounting bond obligations. The OSHD was "flush with money" for highway projects.[16]

In the 1917-18 biennium, a leaderless OSHD bridge department designed ninety-five bridges and fifteen culverts for the state road system, and an additional thirty-four bridges and six culverts for county routes. Construction costs totaled nearly $800,000. The road department paved 50 miles of highway with asphalt or concrete and 112 miles in macadam, and graded 135 miles of roadbeds. The state expended $3.6 million on road and bridge projects during the biennium.[17]

Oregonians viewed the scale of the enterprise and wished it were greater. The 1919 legislative assembly increased the highway bond by $10 million and asked for voter approval of a constitutional amendment raising county obligation limits. It also submitted a referendum calling for a coastal defense route, the Roosevelt Military Highway, as an outgrowth of post-World War I emergency preparedness. Finally, it enacted a one-cent-per-gallon tax on all fuels used in motor vehicles and designated the revenue exclusively for highway improvement. This last measure, Oregon's "gas tax," became the first of its type in the country. North Dakota, New Mexico, and Colorado followed suit, and within ten years, all forty-eight states had adopted similar laws.[18]

The public's overwhelming support of highway improvement measures appeared to assure passage of the June referendum. The OSHD saw this enthusiasm as a mandate to quicken the pace of its road- and bridge-building activities and determined that the state needed a qualified engineer to oversee all aspects of bridge design and construction on state highways and many county roads.[19]

In April 1919, the OSHC chose Conde B. McCullough as state bridge engineer. McCullough accepted the job offer without hesitation. He resigned his professorship at OAC two months before the end of the spring semester and moved to Salem. President Kerr disliked losing one of his most respected faculty, but he recognized the opportunity for McCullough. He accepted the resignation.

Anson Marston, McCullough's mentor and professor at ISC, had written about the essential qualities of engineers. They must be thoroughly trained in the fundamentals and have skill that only practice could yield. Young graduates, he believed, were "not yet engineers ... [because] they became real engineers only after years of successful engineering practice." Their technical qualifications must be of the highest level, and each, he believed, should "give the fellow-members of his profession the benefit of his own experience," through publications and formal presentations. Finally, Marston maintained that a civil engineer was obliged to "have the keenest interest in his work."[20]

McCullough exhibited all of the qualities Marston engendered in his best students. The OSHC recognized this. His accomplishments in Iowa had made him one of the most qualified individuals in the country. He had received high recommendations upon his appointment to the engineering faculty of OAC in 1916, and when he left Corvallis three years later the school newspaper reported that he was "considered the best man in [concrete construction] in the country."[21]

Although creating a comprehensive bridge-building program for Oregon posed a new challenge for McCullough, the circumstances were familiar. He had encountered a similar situation in Iowa nearly a decade before, where he had needed a thorough understanding of the state's varied topography in order to select an appropriate bridge type for each site. He also faced the same dilemma of creating both cost-efficient and aesthetically pleasing bridges. In Oregon the range of topography was greater and so were the design challenges. Purcell's assistants, Lewis W. Metzger and Karl P. Billner, had already created several bridges on the Columbia River Highway with a variety of well-executed structures, many of them delicate, reinforced-concrete arches.[22]

In April, the thirty-two-year-old bridge engineer moved his wife, Marie, and five-year-old son, John, to Salem. He set up offices at the OSHD, which encompassed the entire south wing of the State Capitol's third floor. He assembled a staff of engineers, both as designers in the Salem office, and as construction and maintenance supervisors in districts throughout the state. McCullough's predecessors had relied on a tiny staff to draft plans and supervise projects. But he hired a large staff, creating a drafting office with designers, a general bridge office with contract specialists and statisticians, and a field staff with resident engineers and their assistants.[23]

To fill his personnel needs, McCullough "liberated" the entire senior structural engineering class from OAC to work in his "bridge department." He knew well the capabilities of these young men because he had trained them for three years. President Kerr and other OAC

LATOURELL CREEK BRIDGE
1914

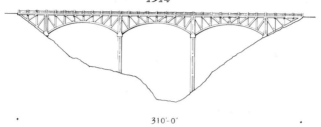

310'-0"

SHEPPERD'S DELL BRIDGE
1914

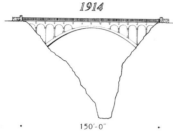

150'-0"

MOFFETT CREEK BRIDGE
1915

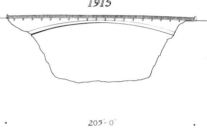

205'-0"

MOSIER CREEK BRIDGE
1920

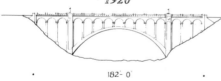

182'-0"

DRY CANYON CREEK BRIDGE
1921

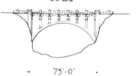

75'-0"

MILL CREEK BRIDGE
1920

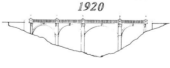

124'-0"

Columbia River Highway Bridges.
(Oregon Historic Bridges Recording Project, HAER,
National Park Service, Todd A. Croteau *et al.*, delineators, 1990)

officials approved of McCullough's idea, providing that the students would complete their remaining course work for other faculty and return to Corvallis for the June commencement. Four of the five classmates readily accepted McCullough's offer. In a postwar economy with high unemployment, it gave them a two-month jump on jobs for new structural engineers. They saw it as an opportunity to apply all they had learned as undergraduates.[24] Ellsworth G. "Rick" Ricketts, P. Mervyn "Steve" Stephenson, Albert G. "Al" Skelton, and Raymond "Peany" Archibald—were all in their early twenties. McCullough assigned them to district offices throughout the state, where they gained practical experience inspecting construction and maintenance on a variety of projects.

But McCullough required additional engineers, some a bit more qualified, to round out his Salem staff. He looked first to OAC and then to Iowa, where he convinced several of his former classmates, with like views about bridge construction, to join him.[25] The first of these had moved to Oregon during the previous year. Merle Rosencrans, an ISC civil engineering graduate, came to Corvallis in 1918, where he taught surveying and mapping as one of McCullough's colleagues in the OAC's civil engineering program. He became McCullough's assistant engineer in the OSHD bridge department. Orrin C. Chase, a member of McCullough's ISC graduating class of 1910, later worked for the ISHC as one of his bridge designers. Initially, Chase served that role again in Oregon, with occasional duties as resident engineer on bridge construction projects in the Salem and Portland districts. Edward S. Thayer also came to Salem from Iowa, and became one of McCullough's chief designers. Finally, William A. Reeves, a 1914 ISC graduate and ISHC employee, became McCullough's "office engineer" in Salem. He assembled plans and specifications for each project, compiled cost estimates and reconciled expenses, and assembled statistical reports on bridge department activities. He helped prepare every OSHD bridge contract. Additional local men filled other vacancies.[26]

In 1919, as McCullough assembled his team of bridge designers and resident engineers, his long-time friend Thomas H. MacDonald succeeded Logan W. Page as director of the U.S. Bureau of Public Roads (BPR), formerly the OPR. MacDonald was the logical replacement. He had served fifteen years as highway engineer for Iowa, and had overseen the creation of the state's comprehensive highway construction and maintenance program. MacDonald's Iowa staff, including McCullough, had advised county governments on adopting specifications standards for roads and bridges, copying the cooperative tone that Logan Page believed yielded better highways for the public.[27] MacDonald and his staff had gained the respect of their counterparts in other states through their membership in the American Association of State Highway Officials (AASHO). As the first president of the Mississippi Valley Conference of State Highway Departments, a regional organization associated with the AASHO, MacDonald had garnered support in 1915 for an innovative federal cosponsorship program. Congress subsequently used his plan as the model for the Federal-Aid Road Act created one year later. His strong good roads beliefs won him membership on the AASHO's executive committee. The AASHO unanimously recommended him to President Woodrow Wilson as Page's successor as director of the BPR. MacDonald was sworn in on 3 May 1919.[28]

A recent historian wrote that MacDonald believed in forming a "partnership between the states and the BPR." He also de-emphasized the moral crusades of his predecessor, instead focusing on simple economic and technical matters. States needed to "approach the task of actual road building in a responsible, sane spirit," MacDonald argued, "that will result in the production of roads rapidly but without extravagance and without loss of faith on the part of the taxpayers." The BPR and state highway departments were no longer simply pulling farmers out of the mud. Instead they were creating a sophisticated, long-lasting, ever-improving primary road system for a much wider public—the American people.[29]

Within his first six years as bridge engineer with the OSHD, McCullough's small department designed nearly six hundred bridges at a cost of $6.4 million. Most were relatively short reinforced-concrete deck-girder spans. Initially, McCullough's goal was to provide structures for as many of the smaller streambeds as possible, advancing the OSHD's goals of grading and paving as many segments of the state's trunk routes as feasible. The number of bridges completed in Oregon in the early 1920s decreased with each year, but the mean length grew. So did the cost. The cost per span in the 1921-22 biennium averaged $8,000; this increased to nearly $12,000 by 1923-24.[30]

As he had learned in Iowa, McCullough found that the best bridge type for a site involved many factors. He argued that the economics of bridge building was "unquestionably the highest, most difficult and most important feature of bridge engineering," because it was "the very corner stone [*sic*] of economy." McCullough did not want to waste taxpayers' money. Accordingly, he paid a great deal of attention to stream behavior, navigational requirements, traffic considerations, architectural features, and state funding.[31]

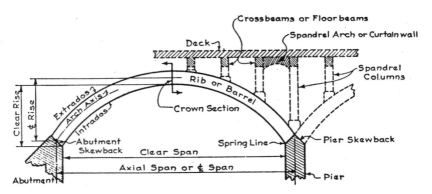

Reinforced concrete arch in elevation. (© Conde B. McCullough 1931, reproduced by permission of John P. McCullough)

Stream characteristics at a crossing, for example, determine the length of a proposed bridge and the types of approaches, abutments, and anchorage it requires. Reinforced–concrete arch bridges need firm foundations. Even the slightest uneven settling of piers or bents can seriously over-stress arch rings, causing possible structural failure. In addition, McCullough considered sight distance, traffic movement, and traffic density in arriving at a final design. The structural members of a wooden covered bridge or a steel through truss, McCullough warned, could easily obstruct driver vision on sharply curved roadway approaches. Routes with truck or commuter traffic require bridges with large carrying capacities. Foresight, McCullough believed, was fundamental to determining if a design accommodated existing and future traffic needs in terms of density, load factors, and width.[32]

He also promoted the idea that "architectural features and scenic considerations," as he called them, dictated the selection of bridge type. In descending order, he ranked bridge categories by their "architectural possibilities" : masonry arches, reinforced–concrete deck spans, steel-plate girder spans with concrete decks and railings, reinforced–concrete girder spans, steel-truss spans, and timber bridges. McCullough's schooling at ISC, his work with James Marsh and the "rainbow arch," and his personal preferences, based on experience, all influenced his ranking of bridge types. He advanced these concepts during his teaching years in Corvallis, influencing the thinking of his students and subsequent employees.[33]

The route of a road in relation to a bridge site, McCullough believed, should also influence type selection. "If the alignment is such that the structure is plainly visible in side elevation from the approaching highway," he explained, "more attention should naturally be paid to [one] which gives a pleasing side elevation outline than if only the roadway were ordinarily visible." Similarly, he also accounted for a bridge's proximity to parks, vacation resorts, and other leisure activity locations. Increasingly, tourism brought revenue to Oregon. Well-maintained highways, with pleasing bridges, drew more tourists, who increased the consumption of goods and services and contributed to the all-important gasoline tax receipts.[34]

Finally, the choice among design types hinged on addressing total costs, both construction and maintenance. McCullough found that bridge types in the top categories of his hierarchy had lower long-term maintenance expenses, but initial construction costs were often higher. He believed that in the long run high-priced maintenance and shorter life span made the least expensive bridge as costly as, or more costly than, the most expensive bridge. "Perhaps the selection of a cheaper construction type," he wondered "may be false economy."[35] A timber trestle span of relatively inexpensive, untreated, and locally cut wood might have a life span of ten to twenty years. A more expensive reinforced-concrete arch span for the same crossing, requiring less maintenance, might last forty to eighty years. Steel-truss spans cost less than reinforced-concrete structures, but they had higher long-term maintenance costs because of their shorter life span of thirty-five to sixty-five years.[36]

Three examples of large spans illustrate McCullough's philosophy of bridge type selection. The Rock Point Arch on the Pacific Highway, in Jackson County, completed in 1920, was McCullough's first large reinforced-concrete structure in Oregon. It replaced a timber-covered toll bridge built in the 1880s as part of an unsuccessful scheme to persuade the Oregon and California Railroad to build a passenger station there. The bridge was on the main stage route through the Sams Valley in southwestern Oregon, which in 1913 was incorporated into the Pacific Highway, one of the OHSC's original trunk routes.[37]

During World War I, the highway department decided that the old covered bridge at Rock Point was inadequate for heavier, wider, and more frequent traffic. Jackson County commissioners promptly requested state assistance in replacing the deteriorating structure. Shortly after his arrival in Salem in the spring of 1919, McCullough designed a reinforced-concrete ribbed deck arch for the Rock Point crossing of the Rogue River. He chose that type for its economy, but also because topography provided "a wonderful setting for a beautiful structure. . . ."[38]

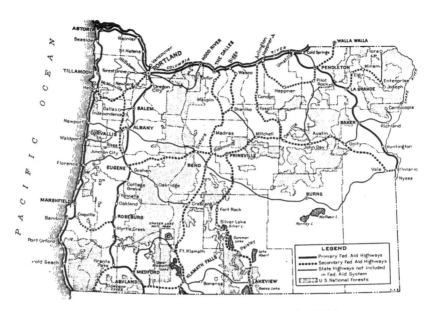

Federal-Aid Highways in Oregon, early 1920s.
(Oregon Department of Transportation)

The 442-foot-long bridge included a 113-foot reinforced-concrete ribbed deck arch between a series of 25-foot reinforced-concrete deck girder spans. The roadway measured 19 feet. The concrete railings consisted of beveled caps nearly one foot wide atop tall urn-shaped balusters. The main arch spanned a deep gorge, with a streambed more than forty feet below its springline. For reinforced-concrete arches, the cost of constructing massive abutments or piers to withstand the arch ribs' thrust often represents a disproportionately large percentage of the budget. Nevertheless, Rock Point's topography lent itself well to this type of structure. The ribs sprang "from bluff to bluff," McCullough wrote, and the rib thrust was "thrown directly into the natural rock footing." This reduced the overall budget for the bridge.[39]

The Rogue River's deep gorge was the Rock Point site's only drawback. It prevented McCullough from using the traditional trestle-type falsework or centering, in constructing the central arch span. He improvised instead, supporting the forms for the concrete ribs with a pair of wooden Howe trusses. Work crews constructed a raft and attached it to the banks of the stream with heavy wire cables. Next, they assembled the temporary structure, anchored it to the rock walls, and dismantled

Rock Point Arch, 1920.
(Oregon Department of Transportation)

the raft. In effect, they had built a bridge to support the formwork of another. In November 1919, after crews had laid out the reinforcing bar, they poured the concrete.[40]

Workers continually battled the dangers of the Rogue River. Swift currents at Rock Point, where the channel narrows, combined with water levels that could rise as much as six feet in just hours, brought the water level within three inches of the lower chords of the Howe trusses. Even the bravest carpenters and masons feared for their safety. Nevertheless, McCullough's ingenuity prevailed. Work was completed on schedule on 17 February 1920. Costs for the project totaled $48,393.91, six times the average in Oregon. "Since the construction was completed, much favorable comment has been heard," McCullough noted in 1920, "so it is believed that the attempt was successful."[41]

OSHD bridge designers favored the reinforced-concrete deck arch for major long-span structures. McCullough repeated the Rock Point Bridge design in two locations along the Columbia River Highway in Wasco County: over Mosier Creek (1920) and Dry Canyon Creek (1921), he created ribbed deck arches. Both structures harmonized with Billner's and Metzger's 1914 reinforced-concrete bridges on the Multnomah County portion of the Columbia River Highway, including spans over Latourell Creek, Shepperd's Dell, Multnomah Creek, Moffett Creek, and Eagle Creek.

In 1922, the OSHD bridge designers finished plans for a long-awaited span crossing the Willamette River at Oregon City, on the Pacific Highway, southeast of Portland. Completed in 1888, Oregon City's old wooden and cable suspension bridge over the Willamette had outlived its usefulness, and Clackamas County and the OSHD had discussed replacing it since 1917. While his first choice was a reinforced-concrete arch, the location presented difficulties that forced McCullough to choose another bridge type. He studied the site from nearly his first day in Salem and acknowledged numerous problems associated with the location. He rejected a reinforced-concrete deck arch because steady river traffic and the depth of the Willamette's main channel prohibited using wooden falsework to support the forms during construction. Similarly, he discarded the idea of building a steel-truss span or suspension bridge because nearby pulp and paper mills emitted highly corrosive sulfur dioxide gas into the atmosphere, which would damage painted steel plating and twisted wire cables.[42]

McCullough believed that a concrete-covered steel span, with reinforced-concrete piers and approaches, would protect the metal components of the structure. Plans called for a road deck resting partway up the ribs of the central parabolic arch, to provide necessary clearance for shipping. The curved shape would also fit well with the natural beauty of the setting's tree-covered rolling hills. But the design was costly. McCullough instead chose a less expensive version of the structure covered with gunite (a mixture of sand, cement, and water).[43]

The new Oregon City bridge's reinforced-concrete deck girder spans and central 360-foot steel through arch measured 745 feet. McCullough arrived at two similar alternatives for erecting the main span; both required towers, either on the shore or in the channel. Eventually, he arrived at a

Construction of Willamette River Bridge at Oregon City, 1922
(Oregon Department of Transportation)

simpler solution. Workers used portions of the old suspension bridge in assembling the arch's box-steel ribs. Cable stays secured to the old towers anchored sections of the arch ribs near their haunches, and the main suspension cables of the condemned span cradled falsework supporting mid-sections of the arch ring). The ingenuity of McCullough and others greatly reduced the construction costs.[44]

McCullough took great care with the outward appearances of all his bridges. On the structure in Oregon City, he covered the exposed metal surfaces of the arch with a mesh of heavy-gauge wire, and then workers sprayed the wire with a thick coat of gunite. This technique provided the bridge with a smooth plaster-like finish and protected the metal from corrosion, as well as giving it a solid reinforced-concrete appearance and helping to placate local apprehension over the lightly constructed span. As seen on many of his other reinforced-concrete bridges of this period, McCullough had the surfaces of recessed panels pebble-dashed, or bush-hammered. This revealed the underlying aggregate, giving a rough surface that contrasted with the rest of the structure. It was an inexpensive detail that added to the aesthetic composition of many of his bridges.[45]

The Willamette River Bridge at Oregon City was the first gunite-covered box-steel arch bridge in Oregon. Although he would have preferred reinforced-concrete construction, McCullough found no alternative to steel and gunite, given the problems he faced in designing a long-lasting, economical structure that fit well with its surroundings.

Willamette River Bridge at Oregon City, 1922
(Oregon Department of Transportation)

He was pleased with the Oregon City bridge project; in November 1922, he informed an acquaintance that he was "foolishly proud" of comment that the structure received from the trade journal *Engineering News-Record.* Editor E. J. Mehren was impressed with McCullough's "care and study devoted to both the selection of type and method of erection. . . ." A 360-foot arch without falsework, McCullough believed, was "an engineering problem ten to one more difficult than any problem presented by any other span [in the region]." In addition, he judged that it was "not surpassed for quality by any bridge work in the United States." [46]

The Winchester Bridge over the North Umpqua River in Douglas County, completed in 1924, was McCullough's longest reinforced-concrete ribbed deck arch structure of the 1920s. More than the others of this type, it emphasized aesthetic considerations, with strong Tudor and Gothic treatment of spandrel columns, curtain walls, and pedestrian plazas. Its long rolling series of arches exemplified McCullough's view of the "quiet, simple dignity" of the reinforced-concrete arch's lines. [47]

The OSHD replaced several outdated bridges along the Pacific Highway from Portland to the California state line as part of a special federally assisted program to improve the route. In 1922, construction

Winchester Bridge over the North Umpqua River, 1924.
(Oregon Department of Transportation)

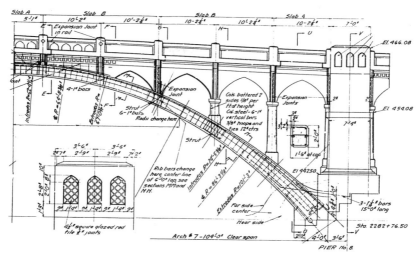

Winchester Bridge over the North Umpqua River.
(Drawing number 1780, Oregon Department of Transportation)

began on an 800-foot seven-panel reinforced-concrete deck arch over the North Umpqua River at Winchester to replace an 1912 steel truss span.[48]

McCullough chose reinforced concrete construction instead of the more traditional steel through-truss design for this large structure for economic reasons. As part of his ongoing study of the economics of bridge design, he concluded that reinforced-concrete arch structures at Winchester and two other locations represented a prudent use of tax dollars. They spanned wide crossings with adequate foundations and connected state trunk routes receiving heavy traffic. He countered the charges of critics that the North Umpqua channel was ill-suited for this type of construction, arguing that problems at Winchester were identical to those encountered at the narrow rock-walled chasm of the Rogue River at Rock Point or Dry Creek Canyon on the Columbia River Highway. He believed that reinforced-concrete bridges easily spanned this type of channel because of its ideal arch anchorages. Moreover, he found that the North Umpqua's riverbed, with its basalt rock base, provided firm a foundation for piers, eliminating the need for expensive coffer dams, wooden piling, or leak-prone concrete caps.[49]

This bridge, more than any other arch structure designed by McCullough in the 1920s, filled its setting with the ambience of Elizabethan England and illustrated his flair for original and artistic design.

He purposely chose Gothic arch-shaped curtain walls and balustrades instead of a commonplace, monotonous, and boxy design. He formalized the structure with Tudor-style observation balconies atop the piers at the road deck entrances—two at each end—and created inset panels for them with red, diamond-shaped inlaid tile. Decorative curved elbow bracketing and bands of dentils below the balustrades added to the structure's architectural design and completed its historical theme. The decoration was relatively inexpensive but greatly enhanced the pleasing side elevations. The railing treatment used at Winchester became one of several stock items. The dentils, balusters, and rail caps were pre-cast, assembled, and applied to the bridge after structural work was complete. These diverse elements formed an integral part of McCullough's philosophy of bridge design and fit well with his ideals of combining economical design with aesthetics.[50]

The costs of the Winchester Bridge over the North Umpqua River totaled $132,759.90. The federal-aid road fund provided approximately $70,000 and Douglas County contributed $60,000. It was truly an innovative venture involving federal, state, and local cooperation. The costs were competitive with those for a simple steel-truss bridge, but it's the bridge's longer life span and lower annual maintenance costs made it more economical.[51]

⊷⊷

CONDE B. McCULLOUGH'S EARLY YEARS in Oregon marked a time of growth both for him and the OSHD. During his brief career as an educator at OAC, he revealed to his students the real possibility of creating economical bridges that were also both structurally sound and aesthetically pleasing. He continued professing his views and educating the public by building bridges that epitomized his philosophy of design. Whether a small wooden covered bridge on a graveled byway or a multi-span structure serving a major highway, McCullough's bridges served rational, straightforward, and practical public needs.

During his early years in Oregon, McCullough had become a respected structural engineer of local, regional, and national dimensions. Though an individualist, he was also a team player in Oregon's and the nation's quest for improved intrastate and interstate highways. The last half of the 1920s would bring him and his adopted state greater recognition as progressive and innovative leaders in highway bridge design.

❧ 5 ❧

Achieving a Reputation for the State of Oregon, 1925-32

A bridge is like a house. Each bridge and each house is a special case; each must be shaped according to the environment with which it is to cope and the function it is to have.[1]

Siegfried Giedion
Space, Time and Architecture

THE YEARS FROM 1925 TO 1932 MARKED a period in Oregon highway history that demonstrated the state's deep commitment to well-built local and national highway systems. For Conde B. McCullough, they represented his ascent to prominence as designer, author, and researcher of roads and bridges. He promoted the economics of highway bridge types through a textbook written for engineering students and practitioners. His continued interest in litigation involving road and bridge construction projects led him to acquire a law degree in 1928. McCullough's willingness to take on challenges of great complexity prompted him to investigate obscure design and construction techniques. He used the Freyssinet method for arch rib precompression on a bridge spanning the mouth of the Rogue River, creating an economically sound structure and simultaneously advancing the field of study. His successes during these years contributed to the greater knowledge of engineering in principle and practice. Both he and the state of Oregon received numerous accolades for these endeavors.

⊱⊰

THE SPRING OF 1925 MARKED THE SEVENTH anniversary of McCullough's tenure as bridge engineer for the Oregon State Highway Department (OSHD). It was his tenth year as an Oregonian and his fifteenth as a practicing engineer. He had seen many advances in his field, including the formation of sound highway department organizations and the creation of a federal highway assistance program directed by the U. S. Bureau of Public Roads (BPR).

McCullough's staff by 1925 numbered twenty-five. He continued to hire numerous recruits from Oregon Agricultural College (OAC) and Iowa State College. Of the OAC graduates, Merle Rosencrans was assistant bridge engineer, Glenn S. Paxson was field engineer, and William A. Reeves was one of three designers. Albert G. Skelton, Orrin C. Chase, and Skelton's younger brother, J. T., another OAC graduate, were three of his five resident engineers located at district offices. Ellsworth G. Ricketts and Mervyn Stephenson were superintendents of bridge maintenance for state routes. The staff also included several locally hired bridge foremen and three drawbridge operators.[2]

McCullough continued his aggressive bridge-building program. In the 1925–26 biennium, his department designed forty-six large structures (greater than twenty feet in length), seventy-three small spans (less than twenty feet in length), and forty-two bridges for county routes. By 1927–28, the figures were twenty-three, twenty-five, and thirty-five, respectively. In 1929–30, the staff drafted plans for fifty-five large-span and seventeen small-span bridges. By 1931–32, these two categories included a total of sixty-eight bridges. A breakdown of figures for this last biennium was not published, but the percentage of OSHD-designed small spans continued to decrease. McCullough and his staff were continuing to implement the OSHD's philosophy of completing as many continuous segments of the state's primary and secondary highway system as possible for speedy delivery of mail, market crops, and other goods. Consequently, they had at first designed and built scores of small spans, together with a few larger structures, thereby extending stretches of passable roads. As the number of proposed new crossings decreased, the bridge department devoted more time to designing larger spans—some new crossings, and some replacements for aging structures.[3]

Thomas H. MacDonald, chief of the BPR and close personal friend of McCullough, had sought supplemental Federal-Aid Road Act legislation in 1921 to strengthen state highway departments across the country, hoping to displace counties as the primary governmental organizations responsible for road and bridge planning and construction. MacDonald argued in 1925 that "the bulwarks of the [federal] system of highway administration are state highway departments." These state agencies had established standards for road and bridge construction that were unprecedented.[4]

Initial construction expenses for spans on market or county roads, paid for entirely by the state or county, dictated the prevailing types of structure on this large segment of the road system within Oregon. Here, traffic density was much less than on routes on the national highway systems, and these county bridges were constructed to carry lighter loads, with road decks often limited to one-way traffic. McCullough pursued a program of building inexpensive, high-quality spans for these routes. Timber construction, though having a limited life expectancy, was the most economical. If traffic demands subsequently warranted new bridges, it would cost far less to upgrade them than to initially construct needlessly elaborate structures.[5]

At many locations, a simple truss-type design adequately spanned crossings of over twenty feet in length. The cheapest bridge was constructed almost entirely of untreated timbers. McCullough created simple modifications that increased its life: for example, the addition of cast-iron shoes at joints eliminated decay near the ends of the wooden members and extended their life span. He justified the added cost of 15 percent for the iron components only if he projected the need to extend a bridge's service life.[6]

Another way to extend the life of a timber bridge was the housed timber structure or covered bridge that had been a part of Americana since the late eighteenth century. Covered bridges had crossed streams throughout the Willamette Valley and other locations in western Oregon since the 1850s. An efficient and economical Howe truss configuration provided adequate live-load capacity for most secondary routes, but without its characteristic covering of gabled roof and shiplap or board-and-batten siding, moisture easily penetrated and decayed the load-bearing members of the truss, causing premature failure. Enclosing a wooden bridge increased the structural components' life span by as much as 150 percent, and it cost far less than replacing the bridge.[7]

McCullough once complained that the nostalgic covered bridge of yesteryear was "an unsightly thing and the modern housed timber truss must not be confused with it." He designed a modern version of the

Grave Creek Bridge, Josephine County.
(James B. Norman, Oregon Department of Transportation)

venerable covered bridge that met the standards for minimum live-load capacities and roadway widths established by the American Association of State Highway Officials (AASHO) in the mid-1920s. In addition, he eliminated the obvious flaws in the old-style spans, such as their saggy, unkempt, and often crude appearance, and stylized them with curved portal entryways and better-proportioned windows. His bridge department created a file of standardized plans from which many western Oregon counties constructed scores of wooden structures.[8]

After devoting his first seven years as state bridge engineer to creating an improved bridge construction program for the state's roadways, McCullough focused on the state's primary highway system, which consisted of major routes such as the Pacific Highway (U.S. 99), the Columbia River Highway and the Old Oregon Trail Highway (U.S. 30), the The Dalles—California Highway (U.S. 197 and 97), and the Roosevelt Coast Military Highway (U.S. 101). He continued his work of selecting the appropriate bridge type for each proposed location. Most reinforced-concrete arched bridges were located in the valleys of western and southern Oregon, over streams emptying into the Columbia River, along the coast, and in the mountainous northeastern corner of the state. The criteria of suitable foundations, climatic conditions, aesthetics, and economics continued to play important roles in type selection.[9]

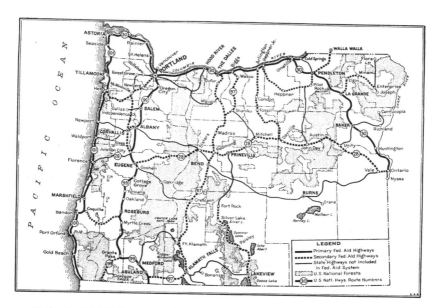

Federal-Aid Highways and U.S. National Highway Routes in Oregon, 1926. (Oregon Department of Transportation)

After their monumental span over the North Umpqua River at Winchester, McCullough and his bridge department designed several large arched structures for the state's heavily traveled routes. The first of these, completed in 1925, was a 134-foot open-spandrel reinforced-concrete deck arch over the Grande Ronde River, located on the Old Oregon Trail Highway, in Union County, near La Grande. With girder approach spans, over adjacent railway trackage, the bridge was more than 300 feet long. A similar 110-foot deck arch, over 400 feet long, was built over the Umatilla River, in Umatilla County, also on the Old Oregon Trail Highway.[10]

McCullough's designers next drafted plans for a 143-foot reinforced-concrete open-spandrel barrel-arch bridge for a crossing on the Pacific Highway over the Rogue River near Gold Hill, in Jackson County. It was the first and only example of this type of structure built in Oregon. In one sense, the span was atypical of the design that McCullough preferred in reinforced-concrete arch spans. Virtually all of his deck-arch structures employed the delicate cost-saving open-spandrel ribbed style of construction. But McCullough feared that a ribbed arch might not withstand the battering from the Rogue River at Gold Hill. After long hours of study, he chose the barrel arch because it gave the structure greater lateral stiffness. Nevertheless, the bridge at Gold Hill had much in common with the Grande Ronde and Umatilla structures. All three shared outward similarities characteristic of McCullough's reinforced-concrete arches, including open spandrels with semi-circular arched curtain walls, standardized ornamental railings, and elbow bracketing.[11]

The most challenging of the mid-1920s bridges was the structure on the The Dalles-California Highway spanning the Crooked River, near Terrebonne, in central Oregon. Proponents of the Crooked River crossing believed it would provide a more direct routing for the major north-south highway than the alternatives. But the gorge, which was more than 300 feet wide, had sheer basalt walls rising over 300 feet from the streambed to the volcanic ash desert. The extreme vertical distance made any use of traditional falsework or centering for concrete spans infeasible. McCullough faced the challenge of designing an economical structure that would provide the load-carrying capacities required on the state's primary road system. He finally chose a 330-foot two-hinged steel braced-spandrel deck arch with reinforced-concrete girder span approaches. A concrete road deck sat atop the entire structure.[12]

The design greatly reduced construction costs and permitted completion of the project in little more than one year. McCullough employed hinges, or rotation points, at the arch's skewbacks and at its crown to economize and expedite construction. The traditional fixed arch was the preferred form for its efficiency in carrying live loads, but its design often involved tedious calculations of stress to determine its

Crooked River High Bridge, 1926. (Photograph by author)

proportions. Using rotation points at the skewbacks created a two-hinged arch that focused the load of the structure on two points. Adding a third hinge at the crown of each arch rib made the structure statically determinate. It also assured easier construction. The drawback was that this three-hinged arch was less rigid than the fixed or two-hinged arches. For the Crooked River High Bridge, McCullough used a third hinge during construction, but keyed or fixed it once the structure was erected. He created a span that required less structural reinforcement near the skewbacks than a traditional fixed arch, but with a greater load capacity than a three-hinged arch.[13]

Site work was difficult. Initially, work crews cut twenty-foot-deep seats in the rock walls of the canyon to anchor the arch's skewbacks, or ends. They built a cableway from one bluff to the other to facilitate easy placement of the structural steel members. The bridge took shape when workers riveted it together piece by piece from the walls in a cantilever fashion until both halves joined at the mid-point above the stream.[14]

McCullough immodestly described the Crooked River High Bridge as "the most spectacular [structure] on the state highway system." It was one of the highest spans in the U.S. when it was completed in the fall of 1926. On the south edge of the gorge, between the bridge and a railroad span just to the west, the OSHD created the fifty-seven-acre Peter Skene Ogden Park, named after an early fur trader and explorer of the region. A 1925 legislative mandate required the highway commission to acquire right-of-way specifically for waysides and parks; this was the beginning of what became the Oregon Parks and Recreation Department. Ogden Park featured a thousand feet of rock parapet walls and overhanging balconies. From here, road-weary travelers could replenish their vitality by viewing the river far below or looking to the east to marvel at the engineering genius of the McCullough's Crooked River High Bridge.[15]

Correspondence between McCullough and Deschutes County Court Judge Robert W. Sawyer concerning vandalism of this bridge illustrates McCullough's aesthetic concerns and reveals his sense of humor. Evidently, young sweethearts had painted their names in three-foot-high letters and drawn some "primitive artwork" on one of the bridge's concrete approach spans. McCullough described their "rather embryonic attempt at mural art," writing:

> *Among other designs and symbols to be found on the north abutment wall is an impressionistic portrait of a tom cat in rear elevation. Far be it from me to decry or deprecate the humble tom cat who, with the possible exception of Magna Carta, constitutes the very bulwark of our*

civilization. However, this matter was considered at considerable length during the preparation of the design and we came to the conclusion that it would be better, in view of the scenic environment, to omit tom cats from our general decorative scheme, and while we are extremely grateful for cooperation, we would much prefer that those who are seeking to assist us in this regard take up the matter with the office first, in order that our general scheme of mural treatment may be more or less standardized.[16]

The OSHD valued the relationship between the bridge and adjoining Ogden Park. McCullough and his designers paid special attention to the color and texture of the stone parapet walls and gravel paths along the river gorge and the bridge's concrete approach spans, hoping to present a river crossing and wayside park that complemented the high desert environment.[17]

⋈ ⋈

MEANWHILE, THOMAS H. MACDONALD, chief of the BPR, promoted the Highway Research Board (HRB) of the National Research Council as the coordinator of studies by a growing number of adept researchers, including McCullough. MacDonald's philosophy for the HRB was akin to McCullough's on highway bridges: both sought to create the most economical and efficient highway transportation system possible for the American public.[18]

In the early 1920s, shortly after McCullough became Oregon's bridge engineer, he began to write on technical issues surrounding bridge design and construction. While he had an affinity for the reinforced-concrete arch, he never lost sight of the need for greater understanding of other bridge types and cultivated a personal interest in movable span bridges—bascule spans, swing spans, and vertical lift spans. He designed two bascule drawbridges for the northernmost portion of the Roosevelt Coast Military Highway at Astoria and Warrenton in 1920, and consulted on reconstruction of Portland's many movable structures spanning the Willamette River. This versatility drew the attention of colleagues who were creating a textbook series on structural engineering.

McCullough became associate editor of a six-volume civil engineering series that George A. Hool and W. S. Kinne, structural engineering professors at the University of Wisconsin, organized in the early 1920s.

He also wrote several chapters for one of the volumes that focused on movable and long-span steel bridges. His writing was featured with other nationally noted civil engineers, including suspension bridge designer David B. Steinman. McCullough authored a large chapter on bascule draw spans, based on his experience with the two coastal structures and the Portland bridges. He also compiled a four-chapter study of steel arch bridges, focusing heavily on the mathematics needed to calculate load capacities for the various types.[19]

In 1927 and 1928, McCullough published two narrowly focused volumes for the BPR on highway bridge locations and surveys. These formed the basis for his own single-volume text, *Economics of Highway Bridge Types*, released in 1929. He wrote the book because it met "a most urgent need in the field of highway engineering" as no published data existed on the topic. It was a general discussion of the subject that promoted his philosophy without the excessive technical material more familiar to bridge engineers and featured a discussion of "the fundamentals of economic analysis and type selection for ordinary highway bridge structures."[20]

McCullough also directed the text to a student audience. He believed that students often lacked a full understanding of the economics of bridge design. The current curricula ignored the subject, which meant that some students had no idea that the "correct type selection is the very corner stone [*sic*] of economy." He wrote:

> *It is true that [bridge] type selection calls for the exercise of the rarest judgment, tempered by long experience in the design, construction, maintenance and operation of bridges under a variety of conditions and it is also true that as a general rule, nothing but time will give the bridge engineer the maturity of judgment needed. It is quite possible, however, to analyze the problem, to separate it, as it were, into its component parts, to state certain fundamental principles and submit certain data which may aid in forming judgments as to probable first costs, maintenance costs, renewals costs, etc., for the various construction types commonly employed.[21]*

As the 1920s progressed, McCullough turned his attention to other areas of structural design and construction. Engineering and the law had occupied his interest since the early days of his professional career. In several instances, he again, as he had in Iowa, served as an expert on bridge construction for court cases involving alleged infringements of design patents. As the pace of road and bridge building increased, so did the need for full-time legal counsel for the OSHD. In 1919, the state legislature named Joseph M. Devers as first full-time counsel. Securing rights-of-way, preparing or reviewing contracts, and defending the state in damage disputes all required the hand of a legal professional.[22]

The first expression of McCullough's interest in Oregon road laws appeared in 1920 in his correspondence with Devers. This began when McCullough reviewed a current edition of Oregon's statutes "Relating to Roads, Highways, Bridges and Ferries." McCullough recommended to Devers that the regulations be altered to make them more compatible with similar codes in other states. This would help eliminate inefficiency in the bridge construction process, from inception of a project to its completion. For instance, the current law forbade county governments from rejecting the lowest bid for a project even if it were greater than the estimated costs of construction. Commissioners often possessed insufficient knowledge about bridge construction and became easy prey for unscrupulous contractors. Supplying county governments with free bridge plans that provided structures meeting AASHO specifications solved only part of the problem. The OSHD, McCullough argued, must be able to inspect construction proposals to ensure that contractors did not overcharge for their services. Moreover, he believed that counties should have the right to reject even the lowest bids on contracts if they were unexpectedly high.[23]

McCullough wanted statutory regulations to apply to minimum live-load capacities of bridges on public roadways. He called for standardized and systematic inspections following a protocol recently devised by his department, including inspecting foundation conditions for piers and bents, and listing estimated repair or renewal costs. Specifically, McCullough wanted to prohibit any state jurisdiction from constructing bridges that could not safely carry the load of a standard ten-ton truck. He acknowledged that considerable cost was involved, but he believed that this would promote construction of high-quality structures. McCullough also recognized that it would reduce the chance of state or local governments becoming involved in costly litigation stemming from unsafe bridges.[24]

McCullough demonstrated his interest in engineering law by seeking Devers' counsel. Both had begun their careers with the OSHD at the same time, and they became close friends. They shared a common interest in the law and its connection to highway construction. Equally important, they formed a social and intellectual bond that brought them and their families together outside the office.[25]

Continued interest in litigation prompted McCullough to enroll in the Willamette University's School of Law in the fall of 1925. In addition to a strong four-year liberal arts curriculum, Willamette, located in Salem, offered a three-year course of study leading to a bachelor of laws or LL.B. degree. This professional program gave students a sound foundation in the fundamental principles of the law. The proximity of Willamette University to Oregon's Supreme Court gave students the opportunity to attend court sessions and ready access to the court's library. McCullough's curriculum was a blend of textbook reading, lectures, and case law. Instead of attending night school, he fit his busy work schedule around his classes. McCullough earned the LL.B. degree in June 1928, and shortly thereafter was admitted to the Oregon Bar.[26]

McCullough argued that engineers should acquaint themselves with the law; it was not necessary for all engineers to become attorneys, but he hoped to make them better practitioners by seeking a "broader outlook, one capable of more enlightened cooperation with the legal profession." Civil engineers inevitably found themselves involved in litigation, either giving expert witness testimony or preparing trial exhibits. McCullough believed that a basic understanding of contracts, torts, real property, and patents better prepared engineers for public or private sector work. Understanding the technicalities of legal procedures made engineers more effective. "All fields overlap," McCullough argued, and a "profound understanding of any profession [including engineering] . . . requires a knowledge of many others." He believed that there was "no narrower man than the specialist who knows naught outside his specialty." As an engineer and a lawyer, McCullough hoped to live to the fullest extent the role of the "efficient manager." He sought not only to apply the "properties of matter" and "sources of power in nature," through his skills as an engineer, he wished to do so through the "rules of civil conduct"—the law, which he believed was an underlying foundation of society.[27]

❧❧

MEANWHILE, ONE OF MCCULLOUGH'S GREATEST challenges involved historic problems associated with crossing the mouth of the Rogue River on Oregon's south coast at Gold Beach. Since World War I, Oregon legislators had supported a "motor road" along the state's coastline as part of a national defense system, envisioning its use in preventing foreign invaders from landing on the West Coast. Initially, the coastal highway was ineligible for Federal-Aid Road Act funds, so in 1919 state voters approved a $2.5 million bond obligation, which matched federal military highway funds to finance the shoreline route. Federal money never materialized, however, and the state government's authority to sell construction bonds lapsed.

By the early 1920s, the pleasure-seeking public, clamoring for improved travel between Portland and the beaches, promoted coastal highway construction. Work had begun in 1921 on a new road designed by the state's highway department. Despite having no connection to the national defense network, the route was known as the "Roosevelt Coast Military Highway." Year in and year out, OSHD crews graded and paved stretch after stretch on the four-hundred-mile route. McCullough's initial contributions included numerous short-span bridges and a few larger reinforced-concrete structures. Most notable were a 150-foot deck arch across the mouth of Depoe Bay and a similar structure over Rocky Creek, both in Lincoln County. In 1928, McCullough and his staff created a third, over Soapstone Creek in Clatsop County. All three exhibited features characteristic of his designs: open spandrels with arched curtain walls; paired arch ribs; and pre-cast decorative railings and brackets.[28]

The state highway engineer, Roy A. Klein, stated repeatedly that the coast highway was the "outstanding and most important objective" of Oregon's road-building program. The route extended the full length of Oregon by the early 1930s and bridged many bodies of water, but the road surface was not a continuous ribbon of asphalt or concrete. Travel frequently became a muddy affair, and slow, inefficient ferries operating across two bays and four rivers contributed to traffic delays.[29]

Ferries had crossed the Rogue River between Gold Beach and nearby Wedderburn since the nineteenth century. Beginning in 1927, McCullough's bridge department took over from private concerns ferry operations at Gold Beach and five other points along the Oregon coast (Coos Bay, in Coos County; the Umpqua River, in Douglas County; the Siuslaw River, in Lane County; and Yaquina Bay and Alsea Bay, in Lincoln County). The goal was uninterrupted service and the state operated the ferries nonstop, sixteen hours a day. The ferries transported between eight and thirty-two automobiles per crossing, except for dry periods or high water that closed operations and stranded motorists for hours at a time.[30]

Private citizens and chambers of commerce along the coast complained bitterly about the slow, unreliable ferry service at the two bays and four estuaries. One critic labeled the crossing between Gold Beach and Wedderburn an abomination. He lamented that the highway would "never amount to anything until there is a bridge built [at Gold Beach]." California vacationers bypassed the Oregon coast, he charged, using instead the Redwood Highway (U.S. 199) and the Pacific Highway (U.S. 99). Unreliable ferry service was depriving the region of tourist dollars it deserved. In 1929, the McMinnville Chamber of Commerce launched a plan calling for reissuing old construction bonds to finance the completion of the Roosevelt Coast Military Highway. Replacing the antiquated ferry system at the six crossings with bridges was the only way Oregon could benefit from the $11 million already spent on the coast highway.[31]

State officials finally called for constructing at Gold Beach what was projected to become the largest Pacific coast bridge between San Francisco and Astoria, Oregon. By June 1929, the state highway commission began studying eight proposals that McCullough and his designers had prepared for a Rogue River crossing. The commission favored a $628,000 structure designed to serve traffic needs for many decades. The bridge included a wide road deck and ample sidewalks and required minimal realignment of existing approaches. That fall, the U. S. War Department's Army Corps of Engineers, which oversaw construction of structures over navigable waters, granted the OSHD a permit to build the bridge.[32]

In early 1930, Congress uncharacteristically pushed for construction as soon as possible. The nation faced crippling economic problems in the wake of the stock market crash. Congressmen saw this and other projects as an integral part of President Herbert Hoover's policy of limited public works aimed at curbing nationwide unemployment. Contributions through Federal-Aid Road Act funds amounted to two-thirds of the bridge's cost.[33]

Characteristically, McCullough mixed aesthetic and practical considerations in the design of the 1,898-foot bridge. He created a multi-arched structure that harmonized with the rolling hills of the coastal mountains and resembled his other large arch bridges. He hoped to economize by employing a relatively obscure decentering technique for reinforced-concrete deck arches, one never previously used in the United States. The method that renowned French bridge engineer Eugène Freyssinet had perfected for shortening ribbed arches had the potential to lower costs by 10 percent over traditional construction methods.[34]

Isaac Lee Patterson Bridge
(Rogue River Bridge at Gold Beach), 1932.

Freyssinet had discovered in the early 1900s that a traditional reinforced–concrete structure with a relatively flat deck arch experienced elastic shortening after its falsework was removed. The span assumed a shorter curve under its own dead load because of axial thrust. Temperature decreases and gradual material shrinkage only accentuated this condition, causing the arch to sag to a lower position. The change in the arch rib's curvature set up tensile, compressive, and deformation stresses, which weakened the structure at its skewbacks. It was prohibitively costly to strengthen arches and piers with additional concrete and reinforcing steel.[35]

As an alternative, Freyssinet introduced a system of hydraulic jacks into the open crowns of the ribs. This technique lengthened the axis of each arch rib by an amount calculated to equal the deformation. Once jacked into position, ribs were keyed with high-strength concrete, and reinforcing bars were spliced at the crowns. Freyssinet theorized that his technique could shorten arch rings to their initial axis without bending. The ribs then carried their own dead load without extraordinary stresses induced at the skewbacks. The method, Freyssinet argued, made possible construction of large-scale arches using slender ribs. From his perspective, McCullough saw the technique as reducing construction costs and saving public funds.[36]

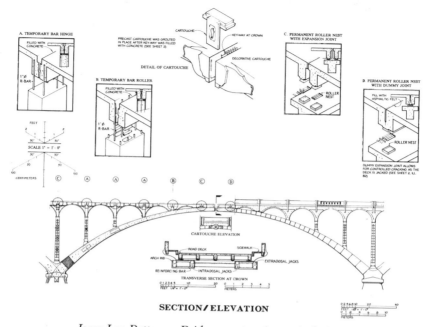

Isaac Lee Patterson Bridge: construction articulations.
(Oregon Historic Bridges Recording Project, HAER, National Park
Service, Todd A. Croteau *et al.*, delineators, 1990)

The structure at Gold Beach was named the Isaac Lee Patterson Bridge
after the Oregon governor who promoted its construction. It is more
widely known, however, as the "Rogue River Bridge." The multi-arched
span consists of a set of nine 16-foot reinforced-concrete deck girder
sections with semicircular arched curtain walls on either end of a series
of seven 230-foot reinforced-concrete ribbed deck arches. McCullough's
application of the Freyssinet method of arch precompression permitted
the use of slender, even delicate, arch ribs that combined his passion for
a mixture of classical designs with embellishments in the emerging popular
"Art Deco" style. Forms for the arch ribs, curtain walls, and railing panels
were constructed to leave impressions in the concrete resembling mortar
joints around *voussoirs*, or wedge-shaped pieces of masonry. Similar scoring
of flat concrete surfaces also gave an illusion of stone construction.
McCullough used elaborate elbow-like brackets to support the sidewalks
and railings. Entry pylons grew out of the first set of piers at each end of
the structure. For their pedestrian passageways he designed simple

Palladian windows in east and west walls and semicircular arched doorways with imitation *voussoirs*, set off with stylized Egyptian sunbursts. He capped them with small two-tiered obelisks. Finally, all concrete surfaces were rubbed smooth with carborundum abrasives to obscure any imperfections caused by the wooden forms.[37]

McCullough employed the Freyssinet method at Gold Beach as part of an experiment in bridge design jointly sponsored by the BPR and the OSHD, and it upheld the agencies' research mandate to determine the advantages and disadvantages of Freyssinet's technique. Moreover, they sought to better understand arch rib behavior after decentering, when falsework and forms were removed, and how the weight of the road deck and spandrel columns on the arch rings affected the distribution of rib stress. Finally, they hoped to learn to what degree formwork prevented the ribs from moving after decentering. They monitored how temperature changes and shrinkage affected rib concrete from the time when it was first poured until it was fully cured.[38]

McCullough had taken on the project with Thomas MacDonald's encouragement. Members of MacDonald's BPR staff in Washington, D.C., especially senior structural engineer, Albin L. Gemeny, worked closely with McCullough and the bridge section of the OSHD. Gemeny had already completed a great deal of research on hingeless reinforced-concrete arches for the Bureau's division of tests. He and McCullough had previously collaborated on BPR technical bulletins. As a team, one complemented the other.[39]

McCullough secured sixteen 250-metric-ton jacks from the Paris firm Freyssinet had contracted for his three-arch bridge over the Elorn River at Plougastel, France. McCullough required this quantity to simultaneously decenter two panels. He used four jacks on the ribs of each arch, two on the extrados and two intrados at the crown. The cost of acquiring additional jacks, combined with the logistics of pouring the entire structure with hand labor in one session, prevented him from employing the technique on all of the arches simultaneously. For each span, McCullough created an elaborate system of piping, hand-activated plunger pumps, and pressure meters to regulate the hydraulic pressure sent to the jacks during the procedure. He designed structural steel emplacements extending ten feet into the ribs from the arch crowns, thereby transferring the thrust of the jacks into the ribs and preventing them from crushing the concrete.[40]

There were other considerations. McCullough feared that the lifting action created in the ribs might adversely affect the rest of the structure.

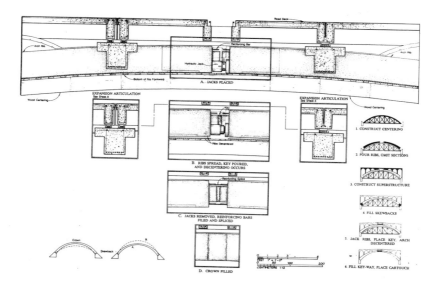

Isaac Lee Patterson Bridge: Freyssinet method of concrete arch construction.
(Oregon Historic Bridges Recording Project, HAER, National Park
Service, Todd A. Croteau *et al.*, delineators, 1990)

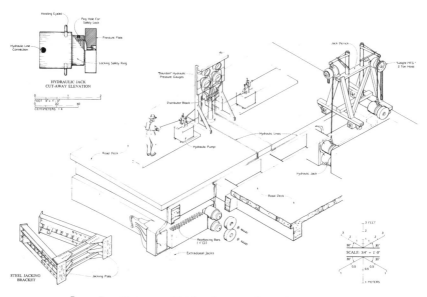

Isaac Lee Patterson Bridge: layout of jacking apparatus.
(Oregon Historic Bridges Recording Project, HAER, National Park
Service, Todd A. Croteau *et al.*, delineators, 1990)

Fixed spandrel columns and the traditionally constructed deck could restrain the ribs and prevent the upward movement, or articulation, he desired during the jacking procedure. But construction schedules necessitated compressing the arch rib sections with the structure, including the deck, in place. Accordingly, McCullough used a complex system of temporary hinges and joints that made the spandrel structure more flexible. This complicated arrangement was the best solution to the problem. Essentially, he had created a floating road deck with spandrel columns that gently rested on the arches and exerted only vertical load forces without creating any additional thrust action.[41]

The construction of the bridge was also a marvel in logistics. In 1930, Gold Beach, Oregon, was remote, some eighty miles from the nearest railway line. Cement and reinforcing bar, pre-cut to proper lengths, were shipped to Port Orford, thirty miles to the north, where trucks conveyed them south to the mouth of the Rogue. Lumber came from local mills, mostly from Bandon, a coastal village sixty miles to the north. Logs for piling came from local forests and concrete aggregates from the river itself. Carpenters built the formwork in an open-air shop adjacent to the construction site.[42]

Once falsework was in place and reinforcing bar arranged, crews poured the concrete for the arch ribs continuously, except for a section near each skewback and at the crowns. The jacks were installed, the Freyssinet technique was applied, and the arches were keyed. McCullough

Isaac Lee Patterson Bridge: view showing falsework during construction, 1931.
(Oregon Department of Transportation)

Isaac Lee Patterson Bridge: close-up of arch crown.
(Oregon Department of Transportation)

recorded the distribution of jack-induced strain throughout the bridge with telemeters and other apparatus placed within the structure. Their measurements revealed how stress was transmitted from ribs to piers in reinforced-concrete deck arch bridges, something Freyssinet had not done.[43]

The French engineer claimed that it was nearly impossible to determine the exact distribution of rib stresses in traditional arches. His method for rib precompression, however, guaranteed a shortening of the arch axis that closely approximated mathematical calculations. Findings collected during the decentering at Gold Beach proved Freyssinet to be correct. His precompression technique caused the arch rings to shorten to their initial point without bending, with the ribs carrying their own dead load without extraordinary stresses induced at the skewbacks. The method eliminated the need for excessive amounts of reinforcing bar and concrete, but McCullough's belief that it reduced total construction costs remained uncertain. Critics contended that expenses for additional skilled laborers needed in the Freyssinet technique equaled or outweighed any savings in materials. In addition, McCullough could not have attempted the project without BPR cooperation.[44]

Isaac Lee Patterson Bridge: close-up of arch ring.
(Oregon Department of Transportation)

The bridge neared completion in late 1931. Contractors originally planned to open the span sometime in January 1932, but an extremely wet December had swollen the Rogue River and disabled the old ferry, the *Rogue*. This prompted an early opening for the bridge, on 24 December 1931. The ferry never ran again at Gold Beach. Crews finally towed the dilapidated vessel north to Waldport where it provided supplemental service transporting vehicles across Alsea Bay.[45]

Completion of the Isaac Lee Patterson Bridge coincided with the OSHC's proclamation that the recently renamed Oregon Coast Highway was complete. Now motorists traveled on a combination of paved and graveled surfaces the length of the state and crossed minor streams by bridge. The popularity of the scenic route prompted speculation that the state would soon replace ferry service at Coos Bay, Reedsport, Florence, Newport, and Waldport with structures similar to the Patterson Bridge.[46]

In April 1932, local citizens, headed by Robert L. Withrow, editor of the *Curry County Reporter* and secretary of the Gold Beach chamber of commerce, began planning a one-day event marking completion of the bridge and the Oregon Coast Highway. Oregon's U.S. Senator Frederick Steiwer asked President Herbert Hoover to participate in the ceremony by operating a historic gold telegraph, which would transmit a signal, relayed from station to station across the country, unlocking a barricade erected across the bridge. The local committee planned a short program with prominent state and regional officials, followed by a luncheon. They organized motorboat races, band concerts, and a dance for the expected crowds of five to ten thousand. The *Coos Bay Times'* banner headlines boasted the "Oregon Coast Highway Now Open For World Travel."[47]

Dedication day went smoothly, except that no one knew if President Hoover had personally activated the telegraph key at the White House. Newspapers reported that because the president was vacationing at Camp Rapidan, Virginia, over the Memorial Day weekend, Vice President Charles Curtis had substituted at the key and "signified to the world that the Rogue River Bridge was a complete structure and open for travel."[48]

In any case, thousands traveled to Gold Beach for the celebration. Many came just to travel over the coast highway, where the ocean was in sight for nearly one-third of the four-hundred-mile route. "Lengthy stretches of open slopes and coves resplendent in native flowers and a luxuriant growth of ferns," and "myriads of shady dells and vistas of ocean," the *Curry County Reporter* intoned, "greet the traveler." Some sought out Gold Beach merely to touch the sixty-three-million-pound bridge that was "a mammoth mass of concrete, artistic in design, which blends harmoniously with its rough and rugged background." The Salem *Oregon Statesman* reported the bridge to be "an interesting structure, a monolithic monument to the design of engineers and the skill of mechanics." Further, it hoped that "All of Oregon will take pride in this bridge. It is an Oregon product and will stand for centuries, we trust, in token of the vision and the courage of the people of this generation."[49]

Conde B. McCullough's experiment, the Isaac Lee Patterson Bridge, was the first and only one of its type constructed in Oregon. No others using hydraulic jacks in the same manner have been built in the United States. McCullough came to see the Freyssinet arch rib precompression technique as more than an economizing measure, though economy remained a goal in all of his designs. More importantly, the experiment contributed significant engineering data regarding the vexing properties of elastic arch bridges. McCullough and Gemeny spoke on the

Conde McCullough, circa 1930.
(Photograph P17-451, Oregon State University Archives)

experiment's results at numerous conferences, authored several professional articles, and wrote a major technical bulletin. McCullough's work at Gold Beach was a genuine contribution to world-class bridge design.[50]

Thus, in less than seven years, McCullough had designed or improved on several bridge types in Oregon, especially the reinforced-concrete arch. He had authored a textbook on highway bridge economics for engineers and students. He had earned a law degree and become an authority on the convergence of civil engineering and the legal profession. He had also received both federal and state cooperation and monetary support for a one-of-a-kind experimental construction project.

During early 1932, when the Patterson bridge project was nearly completed, the OSHD underwent administrative changes that gave McCullough, on paper, a promotion. Roy A. Klein, who had succeeded Herbert Nunn as state highway engineer in mid-1923, retired from the highway department after a nineteen-year career to work for the BPR. McCullough sought the vacant state highway engineer's position, but the state highway commission selected Robert H. Baldock, longtime division engineer at La Grande. As a consolation prize and as a reward for the long hours he had devoted to the Gold Beach bridge, the commission made McCullough the "assistant state highway engineer." It was a kind gesture, but McCullough remained primarily the state bridge engineer, with little, if any, added responsibilities or authority, and no increase in pay.[51]

During the next half decade, McCullough's status as one of the nation's most respected bridge designers made him a regular participant in American Society of Civil Engineers conference sessions. The HRB and AASHO named him a committee member on bridge research and bridge specification standards. He also wrote another textbook. But creating five large bridges for the Oregon Coast Highway, between 1933 and 1936, became McCullough's greatest accomplishment as state bridge engineer.

❧ 6 ❧

The Coast Highway Bridges, 1933-35

Dr. McCullough . . . has become a recognized authority on bridge construction everywhere.[1]

W. E. Emmett
American Institute of Steel Construction
28 November 1934

OST PEOPLE WHO HEAR McCULLOUGH's name will think first of the five major bridges that he designed for the Oregon Coast Highway during the first half of the 1930s. They are the pinnacle of his achievement as a designer, and a lasting monument to his contributions to the state. This was Conde B. McCullough's most prolific period of bridge building, a culmination of years of studying and designing structures. It capped a decade and a half in which he led the Oregon State Highway Commission's (OSHC) publicly mandated mission to build the state's modern highway system. He authored another engineering textbook, received awards for his designing expertise, and constructed several large bridges, including the five major structures to complete the Oregon Coast Highway.

McCullough's second text, *Elastic Arch Bridges*, appeared in 1931, published by John Wiley and Sons of New York. Co-authored with Edward S. Thayer, one of McCullough's chief designers, it covered the mathematical theory of elasticity as applied to designing arch bridges. Like McCullough's previous book, *Economics of Highway Bridge Types*, this volume provided both a reference for professional engineers and a text for undergraduate students. The authors focused primarily on reinforced-concrete structures, but they also discussed steel trussed types of elastic arches. The book featured Oregon's bridges, allowing the authors to draw from their personal experiences.[2]

McCullough argued that the way bridge designers calculated the types and amounts of materials needed to construct an arch bridge had changed greatly since the days when skilled craftsmen created spans from cut stone blocks. Early masons had relied on trial-and-error methods to determine the size and shape of blocks, or *voussoirs*, needed to build an arch bridge. If arranged properly, the *voussoirs* were held in equilibrium (so the bridge would not collapse) under lateral crown thrust. In other words, the blocks maintained the arch form by exerting force against each other in axial pressure, constrained by piers or abutments.[3]

Later designers used a line of pressure theory to analyze arches made not only from stone but wood, cast and wrought iron, steel, and reinforced concrete. But their method did not accurately account for the differing elastic properties of these materials. Other engineers advanced the hypothesis that an arch was a monolithic elastic unit. Those constructed of wood, for instance, deflected, or bent downward, at a rate different from those constructed of iron, even under identical loads. Since the early 1800s, engineers had developed mathematical equations to analyze arch designs, and by the late nineteenth and early twentieth century, when steel and reinforced concrete became ideal structural materials for arched highway bridge construction, elastic theory had become the preferred method used in designing arches.

McCullough and Thayer focused on the types of structures possible with steel and reinforced concrete, especially fixed and hinged arches. They provided a thorough examination based on current literature as well as research for the bridges they had constructed along Oregon's highways. Their volume synthesized the most up-to-date knowledge in a textbook format for students.[4]

Engineering periodicals gave *Elastic Arch Bridges* positive reviews. A recent critic has called it one of the most important American texts ever written on the subject. A lengthy critique in the *Canadian Engineer* by C. R. Young of the University of Toronto lauded the work. "Without a doubt," Young extolled, "this is one of the most valuable structural works that have recently become available." The book explored a complex topic that had received little scrutiny. Young's praise, though, went beyond the book's merits. "Indeed," he continued, "members of the profession regard the State of Oregon as one of the most progressive parts of the world as it concerns the bridge engineer."[5]

✖✖

COMPLETING THE OREGON COAST HIGHWAY, U.S. 101, became one of the state highway commission's primary objectives during the 1930s. At the time he was constructing the Isaac Lee Patterson Bridge at Gold Beach, McCullough also designed several smaller structures along the coast highway and five large spans were constructed on the highway from 1934 to 1936.

Among the bridges on the coast highway, none challenged McCullough's ingenuity more than three small stream crossings, one over the Wilson River in Tillamook County and two others at Big Creek and Ten Mile Creek in Lincoln County. Streambeds at all three locations were nearly identical in width and composition. Their 100-foot-wide channels, with sandy foundations, prevented McCullough from using traditional arches, which required abutment piers to counter lateral thrust. The high water level of all three streams was close to roadway grades, which ruled out alternative reinforced-concrete deck-girder spans. Finally, the harsh coastal environment, with its corrosive salt air, precluded the use of steel-truss spans. Accordingly, McCullough created identical 120-foot tied arches for all three crossings. They were some of the first bridges

Thrust Arch
The tendency of arches is to spread at the base. In a thrust arch, the piers and abutments resist the vertical loading and contain the vertical thrust of the arch.

Tied Arch
In a tied arch, tensile members tie the ends of the arch preventing the spread of the arch horizontally. The abutments resist the vertical loading.

Three-hinged Arch
Concrete arches are designed as continuous arches but constructed as three-hinged arches by placing hinges at the crown and skewbacks. The hinges flex under the high stress placed upon them and allow the arch to adjust itself.

Arch types. (Oregon Historic Bridges Recording Project, HAER, National Park Service, Todd A. Croteau *et al.*, delineators, 1990)

Big Creek Bridge, Lane County.
(James B. Norman, Oregon Department of Transportation)

of this type in the United States and were the first in the Far West. Construction on the Wilson River Bridge began in the fall of 1930 and was completed by June 1931, at a cost of $34,000. The two other bridges were completed by the end of the same year.[6]

The design resembled the tied-arch version of James Marsh's "rainbow" bridge, both in form and function. Unlike traditional fixed through arches, its curved ribs and road deck functioned as an integrated structure, much like an archery bow and string. The road deck—the string—held the outward thrust of the arch ribs—the bow—in compression. The entire superstructure rested atop inexpensive, lightly constructed piers that required little thrust-containing reinforcement. McCullough's design differed little in theory from Marsh's, but it diverged greatly in material composition. For example, he substituted more efficient steel reinforcing bars for Marsh's steel plating and latticework. He promoted economy without compromising modern highway traffic standards by using a hinge, or rotation point, near the top of each arch rib to simplify construction.[7]

For these bridges, McCullough chose French engineer Armand Considère's version of a hinge, consisting of bent reinforcing bar bundled with steel hoops and encased in high-strength concrete. McCullough believed that the Considère hinge functioned more efficiently than other designs. The bending movement in arch ribs induced by concrete shrinkage or dead load was focused on its hour-glass shape and the

Wilson River Bridge, 1931. (Oregon Department of Transportation)

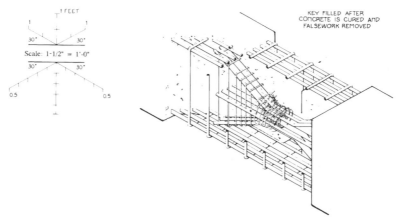

Considère Hinge

This hinge utilizes a concentration of reinforcing bars that channel the loads toward the center of the arch rib. At the narrowest section of the hinge there is a higher steel to concrete ratio, thus making the hinge flexible. Before encasement, the reinforcing steel above and below the hinge is spliced together by welding plates to the reinforcing bars. This arch rib section is then filled with 5,000-psi concrete.

Considère Hinge. (Oregon Historic Bridges Recording Project, HAER, National Park Service, Todd A. Croteau *et al.*, delineators, 1990)

Cape Creek Bridge, Lane County, 1932
(James B. Norman, Oregon Department of Transportation)

hinges, which were immobilized once construction was completed, prevented these forces from weakening the structure. McCullough used the Considère hinge on subsequent bridges.[8]

Other small coast highway bridges posed similar problems. At Cape Creek, twelve miles north of Florence in Lane County, the Oregon State Highway Department (OSHD) planned to realign the road by passing over the stream's deep canyon then tunneling seven hundred feet through Devil's Elbow, a headland, before winding south along the coast. The tunnel and bridge project became known as the "million dollar mile" because of its high cost. The Cape Creek Bridge alone cost $187,500.

McCullough looked at many possible design alternatives before deciding to build a bridge consisting of a reinforced-concrete arch with two-tiered viaduct approaches. The site proved difficult to span because it had steep embankments. Weak foundations prevented the use of rubble fill approaches for an arched span over the stream. In one design proposal he considered using creosote-soaked timber construction for the approach spans, but the BPR, cosponsor of the project, objected, preferring hollow concrete towers and fill. McCullough argued that this design, without

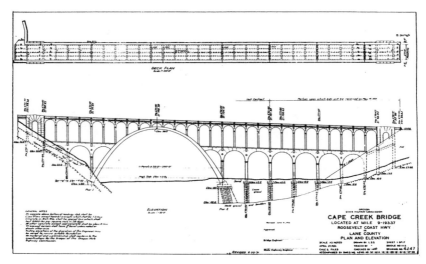

Cape Creek Bridge, 1932.
(Drawing number 4247, Oregon Department of Transportation)

cross-bracing, was susceptible to lateral movement and unsuitable because of the location's unstable substrata.[9]

Eventually, the BPR accepted McCullough's proposal to build a 619-foot, two-tiered viaduct and central arch that mimicked the style of Roman aqueducts, particularly the Pont du Gard, near Nimes, France. The viaduct sections consisted of two-tiered girder span approaches surrounding the 220-foot open-spandrel reinforced-concrete parabolic deck arch, which rose 104 feet over the stream channel. The structure's vertical support members dispersed its load on the unstable foundations, and cross-bracing between the piers and panels prevented lateral movement. Architectural details, including semicircular arched curtain walls on the approaches, elbow brackets supporting the deck, and pre-cast railing panels, were standard elements of structures designed by McCullough. The Cape Creek Bridge was McCullough's and Oregon's only Roman-style concrete viaduct. It was constructed in thirteen months, opening on 30 April 1932.[10]

Optimism abounded with the dedication of the Isaac Lee Patterson Bridge in April 1932 and the Oregon Coast Highway's ceremonial opening in May of the same year, and interest grew in additional construction along the route. Coastal residents and visitors applauded the OSHD spans over Cape Creek, the Wilson River, and numerous other streams and canyons, and they gloried in the miles of asphalt that replaced muddy ruts. An editor of the Portland-based *Oregon Journal* waxed eloquent, stating that "the mountains of the Coast range often thrust themselves sheer into the sea and the new highway turns at times thrillingly to thread their passes." But they expected more. "Three million dollars more must be invested in bridges to replace ferries [at Coos Bay, Reedsport, Florence, Waldport, and Newport]," the *Oregon Journal* inferred, "before the job will be truly done."[11]

By this time, however, the state highway department was suffering from the economic depression that followed the stock market crash of 1929. State highway revenues from various taxes in 1930 had totaled over $12 million, but in 1932 they dropped to just over $9 million. Despite the need for more projects, Oregonians asked for lower, more affordable vehicle license fees, which of course reduced highway improvement funds. The OSHD forecast road construction and maintenance revenues of just $8.6 million for 1933. Federal matching funds remained, but shortfalls in state contributions to joint projects limited dollars available to Oregon.[12]

In the late summer of 1932, Oregonians sought federal help. Congress had recently approved the Relief and Construction Act, which, in part, set aside $120 million in loans for state road construction and maintenance. The lawmakers also created the Reconstruction Finance Corporation (RFC) to loan funds to states for toll bridges and other large-scale projects requiring long-term repayment schedules. The Oregon Coast Highway Association petitioned the OSHD to apply for RFC loans to finance bridge construction at the five remaining ferry crossings on U.S. 101. Local chambers of commerce endorsed the association's plan because it called for building the structures of wood, harvested from local timber stands and cut in local mills.[13]

In the late 1920s and early 1930s, McCullough had conducted a feasibility study of timber or timber-and-concrete composite bridges. In part, he sought a replacement for the wooden covered bridge on routes where, for safety, faster traffic required greater sight distance to bridge approaches and wider spans to accommodate two-way travel.[14]

Even though many western states used timber or timber-composite bridges, McCullough sought to improve upon the general design of these structures for both economic and aesthetic reasons. He saw the need for a rigid grading system to ensure that only the best pressure-treated lumber was used in construction; even if the costs for these materials were greater, their predicted life was much longer. He also found ways to improve upon the structural rigidity of the timber-concrete bridgeto reduce deflection of wooden stringers that caused concrete deck slabs to crack.[15]

In terms of aesthetics, McCullough believed that minor modifications in placement of wooden tower bracing improved the general appearance of the structure. He also found a practical solution to creating aesthetically pleasing details on these low-cost bridges. On the railings, for example, he combined concrete caps and posts with wooden inset panels, creating inexpensive but interesting architectural details.[16]

McCullough endorsed the Coast Highway Association's general plan for replacing the five ferry crossings with bridges. He calculated that sixteen-hour daily ferry service was costing taxpayers $110,000 annually; round-the-clock service that would accommodate projected traffic increases would more than double this figure. With abundant inexpensive building materials, idle labor, and available RFC loans, the state wisely pursued construction plans. A toll of twenty-five cents per vehicle spread over six to ten years, McCullough believed, would generate the $3 million necessary for bridge construction costs. Increased out-of-state tourist travel would brighten prospects even further in the coming years.[17]

The three-member highway commission split on the bridge construction scheme. Its chairman, Leslie M. Scott, believed that tolls might divert traffic away from the coast highway to inland routes, saddling the state with the loan debt and its citizens with higher taxes. The other two members saw bridges, with or without tolls, as a better alternative to the slow and unreliable ferry service. Oregonians from the western third of the state firmly backed the toll bridge plan, but residents from central and eastern sections echoed Scott's fears.[18]

The argument over loan-financed bridge construction collapsed in September 1932, when the RFC declared the proposal ineligible for funding. The state's repayment plan included money from gasoline and vehicle license taxes, which violated the RFC's guidelines for self-supporting projects. Ironically, state law prohibited Oregon from borrowing from the RFC in the first place. For the moment the dream of five new coast highway bridges was dead.[19]

Despite the setback, the Oregon Coast Highway Association remained hopeful that the RFC would provide federal financing for the coastal bridges. In 1933, Joseph M. Devers, legal counsel for the OSHD, researched state law for a way to permit Oregon to legally participate in an RFC loan program, even if it meant creating a private toll group to manage the bridges—but to no avail. Meanwhile, many expected imminent congressional approval of President-elect Franklin Delano Roosevelt's public works legislation, which was expected to distribute $2 to 3 billion for highway projects and other programs. Samuel Dolan, McCullough's former colleague at Oregon Agricultural College and an energetic coast highway promoter, suggested that the state submit a grant application the moment the public works program was enacted.[20]

By May 1933, at Devers' urging, the OSHD began preparing requests for assistance under the soon-to-be created Public Works Administration (PWA). The department asked for a 30 percent outright grant and a 70 percent loan of the estimated $3.4 million in construction costs for five new coastal bridges. Because of the project's labor-intensive nature, Devers and McCullough believed that the proposal was appropriate for PWA funding. They estimated it would employ 750 workers for up to two years and that it would create an additional 375 jobs supplying materials to the construction sites. Both believed that a federally funded multiple-bridge construction project would alleviate the severe economic conditions in the coastal villages and, more generally, the entire state. They also predicted sustained economic growth along an improved Oregon Coast Highway from increased tourist revenues.[21]

As June approached, the state highway commission, determined to build the bridges through RFC loans or PWA loans and grants, called on Oregon's U.S. Senator Charles N. McNary for help. The Navigation Act of 1906 required congressional approval for proposed bridges across shipping channels under War Department jurisdiction. On 11 May 1933, Senator McNary introduced five nearly identical bills authorizing Oregon to "construct, maintain, and operate" the proposed bridges along the Oregon Coast Highway. Purposely, the bills omitted mention of funding except for paying any construction loans from toll revenues. They moved swiftly through Congress, and on 12 June President Roosevelt signed them into law. But bridge construction remained a long way off, because Oregon now needed the War Department's approval for design plans and federal appropriations for construction funds.[22]

McCullough informed the press in early June that his plans were far enough advanced to allow him to begin bridging over Alsea Bay, at Waldport, within thirty days after a federally approved loan agreement. He confidently added that by the end of the summer he could contract construction of the remaining four spans, pending funding..[23]

The Portland *Oregon Journal* reported that State Highway Engineer Robert H. Baldock believed "the bridge structures in question" were the most important that the OSHD had ever undertaken. But he was far less optimistic than McCullough, arguing that the department needed a minimum of six months to finish designing the bridges and begin constructing them. The OSHD needed to offer more than one alternative for War Department approval. "Different schemes," Baldock counseled, "must be tried in order to provide for the public the most suitable and economical structures for each location involved." Officially, the OSHD projected completing bridge designs within three months, since the sooner its plans received final War Department approval, the greater would be its chance for receiving high priority in the competition with other states for PWA loans and grants. Oregon had no other large-scale public works projects even in the planning stages. With its many unemployed, the state had no time to spare on marginal projects. Sheldon F. Sackett, a long-time backer of the coast highway, echoed official views. In his *Coos Bay Times* "Crow's Nest" editorial column, he argued that "bridges are the most feasible proposal the state has available for immediate construction."[24]

McCullough lost no time hiring extra draftsmen for his design team. In its 20 June 1933 issue, the Salem *Capital Journal* reported that capitol custodians were "shocked when a group of bridge engineers appeared for work . . . this morning at 6, earliest hour in memory." McCullough hired more additional draftsmen than he had tables, so that they worked in two shifts, one from 6 A.M. to 3 P.M. and the other from 3 P.M. until midnight.

Since the late 1920s, McCullough had gradually enlarged his bridge department to meet the OSHD's growing highway-building program. He often returned to Oregon Agricultural College (which had been renamed Oregon State Agricultural College in 1927 and was, by the early 1930s, commonly referred to simply as Oregon State College) in the spring, a colleague recalled, to "interview seniors." "With his feet on the chair and casually smoking a cigarette, McCullough would chat with the young men. Somehow the best and the brightest often found their way to the state highway department." One member of the class of 1929, Ivan Merchant, began a long career with the OSHD, starting as a construction inspector on the Isaac Lee Patterson Bridge project at Gold Beach. By 1933, McCullough had promoted him to the design team for the coast bridge project.[25]

In assembling a professional drafting department, McCullough strove to employ designers with considerable architectural training, because they were better able than others to conceptualize design ideas on paper. Edward S. Thayer, Dexter Smith, and Merchant were among

Oregon Coast Highway (U.S. 101) and western Oregon, 1933.
(Oregon Department of Transportation)

McCullough's principal draftsmen. Thayer had been with the bridge department since the early 1920s. Smith had taught mechanical drawing and structural engineering courses at Oregon State Agricultural College until 1929 and came to Salem in 1931 as McCullough's expert in tracing or "inking" final designs onto linen oilcloth. The draftsmen deferred to McCullough's intimate involvement in designing each bridge. "Mac would lay out the overall job," Merchant recalled. "He would pick up a piece of paper and a pencil and say, 'Now . . . this is about what you are going to do.' . . . And he drew this spandrel arch in there and the roadway . . . and there it is. 'Go ahead' [he said]. . . . And about every two or three weeks," Merchant explained, "he'd come back to see how you were getting along." On the lighter side, at the drafting table, P. M. Stephenson recalled that, "We used to have a lot of fun. [McCullough would] come to look at what you were doing. [He'd] look over your desk and see what you were designing and he'd always reach over and get your pencil and start scribbling, stick [the pencil] in his pocket and walk off. So after we got to know him better, as he'd start to move, we'd reach over and grab [the pencil] out of his pocket, never said anything, just reach over and grab it out of his pocket."[26]

The double-shift designers finished plans for bridges over Alsea Bay and the Siuslaw River by August 1933. McCullough expected to complete those for the Umpqua River Bridge by mid-August and the Coos Bay and Yaquina Bay bridges by 1 October. Some thought construction might begin in late fall. But a slow-acting bureaucracy indefinitely delayed federal approval of the Oregon Coast Highway bridge project's $3.4 million PWA loans and grants. The OSHD raised its request to $4 million, citing price increases on construction supplies.[27]

In mid-August, McCullough spoke enthusiastically about the project at a Marshfield Chamber of Commerce luncheon. He praised local citizens and officials for cooperating with the highway commission, and he lauded the coast highway. According to one reporter, McCullough described it as an "unsurpassed advertising medium for the entire state as well as an important transportation link." Oregonians "underestimate the scenic grandeur of the route," McCullough asserted. He reported that engineers and world travelers had informed him that the Oregon Coast Highway was "the most beautiful route in the United States if not the entire world." He saw it as one of the nation's greatest potential tourist attractions.[28]

In mid-September, the OSHD submitted its plans to the PWA's Oregon office in Portland, two weeks ahead of the ninety-day schedule

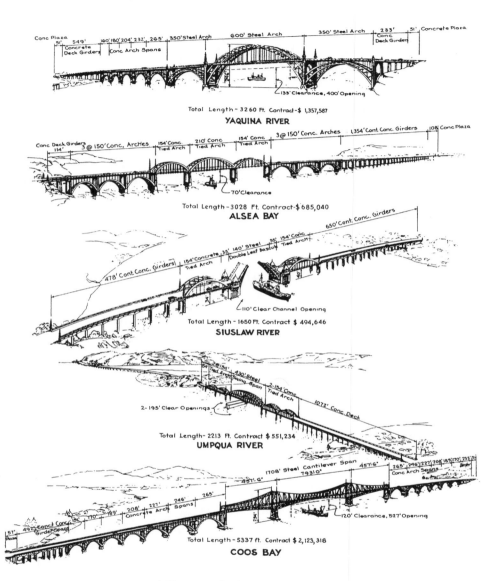

Five coast bridges, October 1933.
(Oregon Department of Transportation)

set in June. Cost estimates for materials and labor had risen once again, this time by 25 percent to $5.1 million. Nevertheless, J. M. Devers, after returning from a meeting in Washington with PWA officials, forecast early construction start-ups. In early April 1934, the PWA approved the coast bridge project's construction plans and financial package.[29]

Meanwhile, in early October 1933, McCullough had quieted public fears that the project might fall through when he unveiled sketches of the proposed structures. For the northernmost bridge spanning Yaquina Bay at Newport, his designers had created a 3,260-foot bridge consisting of a series of reinforced-concrete deck arches rising to a 600-foot steel through-arch flanked by a pair of steel deck arches. For Waldport, they had designed a 3,028-foot structure. It included reinforced-concrete deck-girder and deck-arch approaches to three reinforced-concrete through tied arches over the navigable channel of Alsea Bay.[30]

The Siuslaw and Umpqua rivers supported significant shipping traffic, requiring spans that could accommodate tall sailing vessels. McCullough's designs featured bridges with moveable center sections. For the deep narrow channel of the Siuslaw, his designers created a 1,650-foot structure featuring deck-girder approaches to a central section of two reinforced-concrete tied arches on either side of a double-leaf bascule drawbridge. At Reedsport over the Umpqua's wide and shallow shipping channel, they designed a 2,213-foot bridge that included a steel tied-arch swing span flanked by pairs of reinforced-concrete tied arches.[31]

Finally, at Coos Bay, at Marshfield/North Bend, McCullough and his designers created a 5,337-foot structure including a series of reinforced-concrete deck-arch approaches to a central trussed-steel cantilever section measuring 1,700 feet. The wide bay and the requirements of high shipping clearance made this type of construction the only kind practicable. In addition, crews could easily assemble the 793-foot main section and its adjacent spans without blocking the main channel with falsework.[32]

None of the proposals for the bridges called for wooden construction and this angered some coastal communities and regional lumbermen's associations. The Great Depression brought economic collapse to the Pacific Northwest, with 90 percent of Oregon's timber companies nearly bankrupt. Half or more of the state's timberlands were tax delinquent. Locals objected to building bridges out of steel and concrete when lumber was plentiful and inexpensive.[33]

McCullough reportedly had inflamed the lumbermen in the summer when they learned that an early plan for a wooden bridge over Alsea Bay was only a token effort. He chose instead the designs he released in

October. The editor of the *Pacific Coast Lumber Digest*, C. C. Crow, chastised McCullough as one "who has for many years had a free hand in building these concrete monuments to suit his every fancy." The Salem *Capital Press*, an inveterate opponent of the OSHD, wrote that the state should build wooden bridges and use the remainder of the federal money for improving other portions of the state highway system. It accused McCullough of building "massive concrete monuments . . . for himself at an outrageous cost to the people of the state." While California planned to use 30 million board-feet of Oregon lumber in the San Francisco-Oakland Bay Bridge, the paper continued, "Oregon's arbitrary and pig-headed bridge engineer demands California cement and steel or no bridges."[34]

The atmosphere worsened when a National Lumber Manufacturers' Association representative verbally abused McCullough at an OSHC meeting in Portland. Commission chairman Leslie M. Scott finally banged his gavel and brought the room to order. "I won't have this engineer bullyragged by you or anyone else," Scott announced, and then he explained in great detail how little the state would benefit from building wooden bridges along the coast highway. Bridge lumber would require creosote treating that was only available from suppliers in Washington. First-cost savings for constructing just the approach spans of timber would be insufficient to outweigh the long-term benefits of construction from steel and concrete. Moreover, he noted that there was more than enough work for local timber interests in providing an estimated 10 million board-feet for falsework and forms. Scott continued that the 30 million board-feet ordered for the San Francisco-Oakland Bay Bridge was for falsework and forms, not a wooden bridge. The argument ended when it was revealed that the War Department's Corps of Engineers would not approve plans for large-span bridges constructed of wood over navigable waters.[35]

From late 1933 into 1934, McCullough all but campaigned for the construction of the five coastal bridges. He received numerous invitations for speaking engagements at luncheons and dinners, and he happily accepted them. At one such engagement at North Bend, he revealed his own views of the coast bridge project. The *Coos Bay Times* quoted him as saying that the 400-mile coast highway was the "finest major route in the world." When asked specifically about his proposed bridges, he pictured them not merely as structures carrying traffic, but as "jeweled clasps in a wonderful string of matched pearls."

From October 1933 to April 1934, OSHD officials, coastal communities, and others agonized over what they perceived as federal incompetence, as a communications failure between Washington, D.C. and Portland PWA officials hampered either group from advancing the Oregon bridges project. Harold Ickes vehemently denied accusations that red tape was delaying the vital public works projects in Oregon and other parts of the country. Some believed that the Portland PWA office was concealing something because it remained unwilling to brief the press on the bridge project's status.[36]

In reality, the Portland PWA office, working with field engineers from the BPR, was hastily inspecting hundreds of sheets of construction drawings and work schedules that McCullough and his designer had finished in September. Many of the plans were passed back and forth between Salem's bridge department and the federal offices in Portland until both sides agreed that they met criteria established by the American Association of State Highway Officials (AASHO), the War Department, and the PWA. Federal officials quickly became more skilled in reviewing grant project plans. For example, of the $400 million allotted nationwide for PWA-financed road projects, the BPR had approved contracts totaling only $34 million by September. In just two months, the program dramatically increased its efficiency, with the roads bureau releasing over $236 million for more than four thousand projects employing 276,000 people.[37]

In any event, the Portland PWA office forwarded McCullough's bridge plans to Washington in late October 1933. On 6 January 1934, over two months later, the PWA approved the $5.1 million financial package, with the state receiving 30 percent as an outright grant and 70 percent as a loan secured through bonds and payable through toll revenues. Some coastal communities feared that tolls might have a detrimental effect on the California tourist trade. Others, including the BPR chief Thomas H. MacDonald, objected in principle to tolls on public highways. But other Oregonians, far removed from the coast, argued that statewide automobile owners should not be required to finance the *coast's* bridges through higher license fees and fuel taxes.[38]

The PWA turned the issue over to Oregon lawmakers. The state legislature initially approved issuing special revenue bonds to the federal government bearing 4 percent interest that equaled projected toll collections. An improved bond market, however, persuaded them to issue long-term general obligation bonds free from toll revenue repayment at 2.624 percent interest, thereby saving $1,586,902 on financing costs.[39]

Honorary degree recipients, Oregon State College, 1934.
C. B. McCullough is second from right.
(Neg. 1578, Oregon State University Archives)

On 4 June 1934, William J. Kerr, Chancellor for Higher Education in Oregon and former president of Oregon State College (OSC), presented McCullough with an Honorary Doctor of Engineering degree. He cited McCullough as an "international authority on bridge design." Kerr lauded McCullough for his ability to beautify Oregon with "magnificent examples of modern bridges," which distinguished the state as a "progressive exponent of bridge building." He acknowledged McCullough's membership in "leading honorary and technical societies in engineering" and his authorship of several "books, major reference works and technical articles in the field of bridge engineering" as equally noteworthy achievements. In November, McCullough's John McLoughlin Bridge over the Clackamas River, at Oregon City, received an excellence in design award from the American Institute of Steel Construction. Both these honors vaulted McCullough and Oregon's bridge program into the national spotlight.[40]

McCullough had designed the John McLoughlin Bridge in 1932 as part of the OSHD's expressway project on U.S. 99E, the Pacific Highway East, between Portland and Oregon City. The location with its rolling tree-covered hills lent itself to a bridge with strong architectural features. McCullough's plans called for a 720-foot bridge with a central 240-foot

John McLoughlin Bridge over the Clackamas River, 1933
(Oregon Department of Transportation)

*Plaque commemorating McCullough's excellence in design award from the
American Institute of Steel Construction for the John McLoughlin Bridge*
(Oregon Department of Transportation)

steel tied arch flanked by a pair of similar 140-foot spans. The road deck was wide enough to accommodate four lanes of traffic. Arch-shaped openings in the fluted main piers and ornate Art Deco-style entry pylons served as precursors to the architectural treatment on the large coastal bridges.[41]

What distinguished this structure was its unique tied-arch system. Though similar in design to the three reinforced-concrete bridges at Wilson River, Big Creek, and Ten Mile Creek, it featured steel box girder construction. In addition, McCullough and Orrin C. Chase, his chief designer for this project, created a novel system whereby the tension members of the tied-arch span connected to the arch ribs at both ends of each span by a series of steel eyebars. The rods were arranged below the wearing surface of the deck, through holes cut in the web of the deck beams. Previously, McCullough had used steel reinforcing bar embedded in the concrete deck to connect the arch ribs. His design provided a slight reduction in materials over an alternative simple steel through-truss span.[42]

The tied arch design that McCullough chose used slender, delicate hangers to gracefully suspend the deck from the arch ribs, thereby presenting the motorist a relatively unobstructed view of the landscape. It also afforded side elevation outlines more open than ever thought possible with a steel truss. Engineering aspects of the bridge helped create its architectural merit for which it received the prestigious honor. McCullough would repeat this design in two almost identical three-span steel arch bridges—the South Umpqua River Bridge at Winston, in Douglas County, in 1934; and the Eagle Creek Bridge, on the Columbia River Highway, in Multnomah County, in 1936.[43]

About the time of the OSC commencement, McCullough and the state highway commission learned that the American Institute of Steel Construction (AISC) had designated the McLoughlin Bridge as America's "Most Beautiful" steel bridge constructed in 1933, costing less than $250,000. The Institute had begun the award program six years earlier, placing plaques on structures in various categories based on cost as worthy examples of steel bridge design. In presenting a steel plaque to the OSHD and a medal to McCullough on 28 November 1934, AISC representative W. E. Emmett stated that the bridge stood as "an achievement of the Oregon state highway department [*sic*] and its personnel who have done so much for the advancement of bridge building not only in their own state but throughout the United States." In this same vein, he added, "Dr. McCullough . . . has become a recognized authority on bridge

Oregon Coast Highway Bridges.
(Oregon Historic Bridges Recording Project, HAER, National Park
Service, Todd A. Croteau *et al.*, delineators, 1990)

construction everywhere."[44] By 1934, it was clear that McCullough was
more than an engineer. The public and his professional colleagues
exhibited high regard for him as a master bridge designer who crafted
beautiful bridges, which he sculpted out of concrete and steel.[45]]

Bids were opened for the coastal bridges in the spring of 1934, and
by August all five were under construction. McCullough provided staffs
of field engineers to oversee the progress on each structure and contractors
employed eight hundred men on-site and another seven hundred in
related jobs on the multi-bridge project. Despite the economic depression,
costs for labor and materials increased project estimates to $5.6 million.[46]

Conde B. McCullough's five large coastal bridges represented the
pinnacle of design, both aesthetically and technically. More than his
previous structures, this group of bridges featured the possibilities of

SIUSLAW RIVER BRIDGE ~ 1936
1643 FEET, 500.78 METERS
[HAER NO. OR-10]

478FT · 170FT·157FT·170FT · 650FT

UMPQUA RIVER BRIDGE ~ 1936
2216 FEET, 675.59 METERS
[HAER NO. OR-45]

87FT · 155FT·155FT · 437FT · 155FT·155FT · 1072FT

WILSON RIVER BRIDGE ~ 1931
185 FEET, 56.39 METERS
[HAER NO. OR-39]

30FT·123FT·36FT

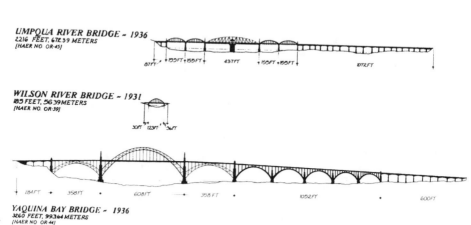

284FT · 358FT · 608FT · 358FT · 1052FT · 600FT

YAQUINA BAY BRIDGE ~ 1936
3260 FEET, 993.64 METERS
[HAER NO. OR-44]

793FT · 458FT · 1320FT · 283FT

reinforced concrete. Because of the plasticity of this medium, designers like McCullough could cast concrete with any desired form, not necessarily mimicking masonry construction. He embellished the structures with Art Deco ornamental pylons and spires and with stylized Gothic piers, spandrel brackets, and arched railing panels. Lines were clean and modern with only a hint of the *voussoirs* of the past. Such attention to aesthetic detail in the mid-1930s was, according to one observer, "an artistic expression of optimism in a period of austerity." Landscaped waysides, approached from grand staircases leading from plazas, united the architectural space of the steel and concrete structures with the sandy shores and timbered hills of the natural settings. The repetitive arch form made the bridges appear as continuations of the undulating coastal mountains.[47]

In terms of advanced engineering concepts, McCullough used the Considère temporary construction hinges that he had employed on tied arches over the Wilson River, Big Creek, and Ten Mile Creek. They were practical forms of technology for constructing the deck arches and tied arches throughout the project. Once again, McCullough employed

Yaquina Bay Bridge at Newport, 1936.

Alsea Bay Bridge at Waldport, 1936.

Alsea Bay Bridge: elevation drawing.
(Oregon Historic Bridges Recording Project, HAER, National Park
Service, Todd A. Croteau *et al.*, delineators, 1990)

techniques whose success had been demonstrated in earlier projects, including vibrating machines that rid freshly poured concrete of air pockets and gave it uniformity. The project was labor intensive to fulfill PWA requirements. Workers used handsaws instead of power saws to cut lumber for falsework and wheelbarrows instead of mechanized buckets to transport wet concrete from mixers to forms.[48]

The Yaquina Bay Bridge at Newport, 155 miles south of Astoria and the Columbia River, consisted of a long series of 150-foot reinforced-concrete ribbed deck arches opening on to three steel spans—two 350-foot deck arches flanking a 600-foot parabolic through arch. The structure continued to an observation plaza on the north end with a grand staircase slowly wrapping itself around the pier abutment and leading to a wayside picnic area below. At each end of the central arch, pairs of long, slender entry pylons in the Art Deco-style extended from massive piers, their height already exaggerated by a stepped-back design and vertical scoring, or fluting. Smaller pylons marked the entrances to the central three-arch section of the bridge. Gothic-arch railing panels and pier bracing complemented the more contemporary architectural style. One observer envisioned the bridge as "arching across the water like a ballerina taking several smallish but impressive leaps, one great soaring, breathtaking leap, followed by a succession of smaller leaps to the opposite bank." It cost $1.3 million.[49]

The Alsea Bay Bridge, 20 miles south of Newport, linked the fishing village of Waldport with wooded cliffs to the north. There, a long string of deck-girder spans broke into a series of three 150-foot ribbed deck

Siuslaw River Bridge at Florence, 1936.

Lewis and Clark River Bridge, Clatsop County.
(James B. Norman, Oregon Department of Transportation)

arches, three 154-foot tied arches, and three more 150-foot ribbed deck arches, all of reinforced concrete. Fluted entry pylons marked the bridge's entrances, and stepped obelisks spired with slender tapered Port Orford cedar tips rose at each end of the group of lattice-canopied tied arches. The deck arches, identical to those on the Yaquina Bay Bridge, rhythmically spanned the spaces between the piers with Gothic-arched openings whose repetitiveness seen from under the deck gave the illusion of grandeur on a cathedral-like scale. Short staircases at the south end of the structure led pedestrians to the long, wide beaches of Waldport. The bridge cost $778,000.[50]

The smallest of the five PWA structures was the Siuslaw River Bridge at Florence, 54 miles south of Waldport. Two 154-foot tied arches, identical to those on the Alsea Bay Bridge, flanked a double-leaf bascule drawbridge. The movable midsection provided a 140-foot horizontal clearance between piers for ocean-going ship traffic. The draw-span was similar to McCullough's 1920 bascule bridges over Youngs Bay near Astoria, and over the Lewis and Clark River near Warrenton. In addition to signature entry pylons on the large coast bridges, McCullough disguised pairs of mechanical sheds, two at each end of the draw, as stylized obelisks. They

Umpqua River Bridge: concrete arches on south side of swing span,
Reedsport, Douglas County.
(Photograph by Jet Lowe, summer 1990. Library of Congress, Prints and Photographs Division, Historic American Engineering Record, Reproduction Number (HAER, ORE, 10-REPO, 1-DLC/P2)

Coos Bay Bridge at Marshfield/North Bend, 1936.

Art Deco motifs on stairway at south abutment of Coos Bay Bridge.
(Photograph by Jet Lowe, summer 1990. Library of Congress, Prints
and Photographs Division, Historic American Engineering Record,
Reproduction Number (HAER, ORE, 6–NOBE, 1-DLC/PP15)

contained the gears and motors that moved the bridge's midsection and provided living quarters for the operator. The bridge cost $527,000.[51]

The simplest of the five structures was the Umpqua River Bridge at Reedsport, 20 miles south of Florence. There, lowlands and a wide shallow shipping channel necessitated using a central swing span, similar to one he had created years before over the Coquille River. Two pairs of the 154-foot reinforced-concrete tied arches flanked a 430-foot steel tied-arch center structure, with a tender's shack nestled in the cross bracing of the arch ribs. The only ornamentation on this bridge, other than the Gothic-arch treatment of the piers and railing panels, was the plain entry pylons marking the entrances to the central spans. The structure cost $581,500.[52]

The southernmost and largest of the bridges was the mile-long structure spanning Coos Bay, 31 miles south of Reedsport. The enormous width of the bay, with its 40-foot-high approaches, enabled McCullough to design a 150-foot vertical shipping clearance at mid-span without a movable structure. A series of thirteen reinforced-concrete deck arches, similar to the approach spans of the Yaquina Bay and Alsea Bay bridges, flanked the 1,700-foot steel-trussed midsection. Two large tower piers, each consisting of 34 tons of steel and concrete, rose 280 feet above the water level and supported the ends of the cantilevered structure.

The bridge bore all the architectural treatments found on the other coastal spans. True to form, McCullough provided spacious flowing staircases with intermediate landings for pedestrian traffic that descended from plazas to grassy areas near the ends of the structure. Here, as with the other bridges, McCullough paid close attention to the piers' shape and form because he knew that park visitors would see them up close. The Coos Bay Bridge cost $2,143,400.[53]

David Plowden, an acclaimed critic of bridge design, labeled the Coos Bay structure "An outstanding example of the large steel cantilever." It was one of the few American bridges of this type to employ curved top and bottom chords, giving maximum shipping clearance and overcoming the "basic disparity between the bridge's steelwork and the concrete arch approaches." But Glenn S. Paxson, acting bridge engineer during its construction, gave a business-as-usual response to questions about its captivating qualities. In 1941, he offered this assessment of the bridge.

I am rather at a loss to list any unusual thing about the Coos Bay Bridge. It can be said that this structure is the longest cantilever span in Oregon. There are, however, many other cantilever spans in the United

States which exceed this in length. We have always felt that the most unusual thing about the structure is the attention to appearance given in design. The usual cantilever structure is far from a thing of beauty, but we feel that in this structure we have at least minimized the unpleasant architectural appearance of the usual cantilever spans.[54]

Another observer drew quite different conclusions.

The beauty of the bridge lies in its graceful symmetry. Its arches, steel spires, and gridwork reveal the sensitivity of its designer to historical tradition. But McCullough adapted those architectural design elements to the natural setting of Coos Bay in a uniquely harmonious way. The bridge gives pattern to space and light and to the incessant flow as naturally as the topmost branches of a Douglas fir or channels of an offshore reef.[55]

David Plowden stated succinctly, "Few later bridges of its type have been as outstanding."[56]

The regional press gave the coast bridge project thorough coverage. Professional and trade journals such as *Engineering News-Record* and *Western Construction News and Highways Builder* featured the bridges in descriptions of design and construction techniques.

In response to the aesthetics, architecture, and natural beauty of the Oregon Coast Highway, the state organized the Travel and Information Division of the OSHD in 1935. Highway commissioners hoped that advertising the state's modern road system might increase tourist traffic, boosting gasoline tax revenue and pumping money into the state's sickly economy. The Division printed 234,000 booklets and maps describing the many virtues of Oregon's scenic beauty, including the bridges of the Columbia River and Oregon Coast highways. It distributed the pamphlets to tourist and travel bureaus, service stations, automobile clubs, and individuals. It also received free advertising, depicting Oregon as a vacationland in major regional newspapers and national magazines. The tourism promotion budget of $48,000 in 1936 was small compared to the $750,000 it helped generate in nonresident gasoline tax revenues. It was estimated that out-of-state visitors spent over $18 million in Oregon during that year. Revenues continued to rise until World War II curtailed leisure travel.[57]

All five bridges were completed by the fall of 1936, within the two-year schedule forecast at the project's onset. The Yaquina Bay Bridge opened for traffic on 6 September 1936. Like the others, except for the

Umpqua River Bridge, which unexpectedly opened without fanfare, the Newport bridge was the subject of a community-wide dedicatory celebration also marking the end to ferry service along the Oregon Coast Highway. On 3 October, state and regional dignitaries attended a day of speeches, luncheons, and parades. Even the Coast Guard cutter *Pulaski*, two Navy destroyers, and a squadron of seaplanes paid visits. One particular noteworthy individual, however, was absent—Conde B. McCullough. He had left the state in October of 1935 on BPR chief Thomas H. MacDonald's instructions, embarking on a fifteen-month assignment in Central America designing bridges for the Inter-American Highway.[58]

McCullough's bridges have sparked widespread praise for their aesthetic treatment. Plowden wrote that his structures represented "the most interesting concentration of concrete bridges in America," adding the qualification that McCullough, like many of his contemporaries, frequently added ornamentation that "marred the beautiful simplicity of his engineering." Others disagreed, believing McCullough had a specific reason for embellishing his bridges with Art Deco–style pylons, obelisks, and piers. He "was much too thoughtful to throw gewgaws onto his work to please others. . . . [They] show too much intelligence for that." Instead, the ornamentation enabled the viewer to "feel not only the awesome presence of nature indicated in the flowing form of the bridge, but the unique, sometimes whimsical power of human intelligence that coexists with nature." In other words, McCullough's bridges did more than connect highways; they united river and stream banks with structures whose symmetry and balance complemented the natural setting.[59]

Shortly after completing the coast bridges, McCullough expressed to a friend a cynical impression of the engineer's role in society that revealed much about his philosophy of bridge building. "If we only knew the truth, the decline of ancient Babylon and the desolation of Sodom and Gomorrah were probably dated from the time when they formed the first engineering society." He hoped that the simple elegance of his bridge designs would help to provide an antidote to the problem.[60]

In the early 1930s, Conde B. McCullough had won widespread acclaim as a researcher and designer of highway bridges. The reviews for his *Elastic Arch Bridges*, the recognition that the McLoughlin Bridge gained, and the honorary doctorate awarded him—all were testimony to his expertise unequalled in the United States. McCullough's greatest challenge was creating five large bridges for the Oregon Coast Highway.

Construction of these cost-efficient but elegant structures provided jobs for hundreds of the state's unemployed, and once completed, helped transform coastal fishing villages into tourist destinations that increased tax revenues and strengthened local economies. In the next decade, McCullough would turn his attention to the international bridge-building world on a federally sponsored project—the Inter-American Highway.

❧ 7 ❧

Central America and Beyond

If we engineers had souls[,] which I doubt, we might have to take to the back roads to keep from blushing every time we see some of the things we have done. But on the other hand, I'm kinda human like the rest of humanity, and I'll admit that there's at least one or two bridges I've had a hand in, and when I look at them, I kinda figure I'll have some alibi when I see St. Peter. Not all of 'em, you understand, but some of 'em did come out so good they make life worth living.[1]

"Conde B. McCullough—Bridges,"
editorial, *Register-Guard* (Eugene),
7 May 1946

KNOWING THAT THE OREGON COAST HIGHWAY bridge project was nearing completion under his staff's guidance, McCullough left Oregon to design several bridges for the United States government-sponsored Inter-American Highway. Working for his old friend Thomas H. MacDonald and the U.S. Bureau of Public Roads (BPR), McCullough created three suspension bridges and a dozen smaller structures as part of the country's commitment to a Pan-American Highway stretching from Alaska to the tip of South America.

The United States committed itself to the Inter-American Highway in the 1930s to promote good will and provide Americans with jobs. Since the late nineteenth century, American businessmen had expressed interest in creating a Pan-American railroad to connect North and South America. It was a grand scheme to promote commercial growth in the Western Hemisphere, but little came of it. By the early 1920s, American companies had invested heavily in Latin American petroleum and natural resources, and automobile travel had become firmly entrenched in American life. Businessmen shifted their attention to the construction of an intercontinental highway that would link Latin America with the United States.

The Pan-American Union was an organization of the twenty-one American republics dedicated to hemispheric understanding and peace. At its Sixth International Conference in 1928, it modified the ambitious Pan-American Highway proposal. A more practical plan was devised which involved building an Inter-American highway connecting the U.S. with Panama, via Mexico and the Central American republics. In 1929, President Herbert Hoover hoped to improve relations with these Latin American countries by asking Congress to cooperate with the Pan-American nations on reconnaissance surveys to establish a proposed route.[2]

From 1930 to 1933, the BPR conducted a survey that designated a 3,250-mile route from Laredo, Texas, to the Panama Canal, focusing on 1,400 miles that passed through Guatemala, Honduras, Nicaragua, Costa Rica, and Panama. The highway's route followed the Pacific mountain slope, occasionally skirting into the uplands to serve remote agricultural sections, and it passed through, or near, large towns and capital cities.[3]

In 1934, supporters asked Congress for $5 million to construct bridges and an unpaved road as part of President Franklin D. Roosevelt's Good Neighbor Policy toward Latin America. Congress, because of Depression constraints, appropriated only $1 million for additional surveys and construction. The BPR and the Central American countries agreed in 1935 to use the reduced appropriation for bridge construction, reasoning that without bridges the route could never be completed. It was a cooperative venture to which the republics supplied building materials, manual labor, and an American-trained staff of local engineers. The U.S. furnished engineering supervision and construction supplies and equipment. The project conformed to Roosevelt's New Deal economic recovery plan because it provided jobs for American steel and cement producers.[4]

Chief MacDonald established a BPR field office in San Jose, Costa Rica, in September 1935. He asked the Oregon State Highway Commission (OSHC) to release McCullough from his state obligations so that he could design and supervise construction of bridges along the highway's proposed route. MacDonald chose McCullough because of his extensive knowledge and experience in creating economic bridges.[5]

Conde B. McCullough, 1936, Central America.
(John P. McCullough Collection)

Friends and acquaintances congratulated McCullough on his unique opportunity, but expressed regret at his sudden departure. On the evening of 8 October at Portland's Multnomah Hotel, over one hundred fifty dignitaries attended a banquet in his honor. According to the Portland *Oregon Journal*, the guest list included engineers from federal, state, and local organizations. Oregon state highway commissioners, both past and present, attended as did members of the Portland press corps.[6]

The state highway commissioners presented McCullough with a plaque bearing an etching of the Isaac Lee Patterson Bridge (Rogue River Bridge) acknowledging his many years of service to the state. The dessert featured an ice sculpture scale model of the Patterson Bridge with ice cream automobiles on its road deck. The Salem *Capital Journal* described McCullough's selection for the Inter-American Highway post as "a fitting and earned reward for ability, integrity, enterprise and hard and faithful work in a field . . . [in] which he has won world wide recognition." Two days later, a large gathering of family and friends dined with McCullough at a Salem restaurant.[7]

McCullough expressed his desire to see through to completion his five coast highway bridges, but he decided to leave the project to his bridge department personnel. His staff quickly reorganized. Ray Archibald, McCullough's supervising engineer at the Coos Bay Bridge site, joined him in Central America in November as his chief assistant and designer. Dexter Smith succeeded Archibald as supervisor of the entire coast bridge project. Glenn S. Paxson, field engineer, temporarily overseeing construction of the coast bridges, became acting bridge engineer. Albert G. Skelton, district bridge maintenance engineer at La Grande, assumed Paxson's position on the coast. Mervyn Stephenson from the La Grande district office became field engineer.[8]

McCullough attended briefings in Washington, D.C., before sailing for Panama. He arrived in San Jose, Costa Rica, shortly thereafter.[9] After careful consideration, he designed three suspension bridges for the highway. They were located over western Panama's Río Chiriquí, near David; southern Honduras's Río Choluteca, near Choluteca; and southeastern Guatemala's Río Tamasalupa, near Asunción Mita. McCullough surveyed each site, studying its geologic formation, and researched the traffic predicted to use it. Several factors made designs common to Oregon bridges impracticable. Central America's rainy season produced annually from 60 to 200 inches of precipitation; Oregon's coast, by comparison, receives an average of 70 to 90 inches a year. Swift high water made arch construction using falsework difficult and dangerous, and the region frequently experienced earthquakes. Because of their low center of gravity and flexibility, McCullough opted for suspension bridges, which were less vulnerable to earthquake damage.[10]

Economics and aesthetics also affected McCullough's design choice. Suspension structures did not require highly trained labor. McCullough could employ local low-wage, semi-skilled and unskilled workers to construct concrete anchorages and masonry piers and assemble steel members. Riveted steel plating used in the superstructures arrived in small light bundles that were easily transported to the remote construction sites. Moreover, given the region's climatic and geologic characteristics, McCullough believed the only alternative to the suspension structure would have been a special type of cantilevered span. He decided against it because of its higher cost and "unsightly appearance."[11]

Even though all three bridges had many similarities, each design was unique because of different structural and architectural elements. McCullough constructed the bridges simultaneously, beginning with the span for the Río Chiriquí crossing in Panama. There, he created a

Río Chiriquí Bridge, Inter-American Highway, Panama, 1936.
(Federal Highway Administration)

400-foot cable suspension bridge with unloaded backstays. Its total length, including short enclosed girder approaches to the main span, measured 655 feet. Fifty-eight-foot box-steel towers rose from concrete piers resting on the stream banks' natural rock foundations.[12]

McCullough used bracing between the towers that provided both structural rigidity and low wind resistance. As with his Oregon bridges, he valued architectural aesthetics. He combined medieval details to produce what he labeled a "Spanish Colonial" motif. The tower bracing's bottom chord formed a Gothic arch opening. Above this, McCullough created a bracing panel from varied but symmetrically placed openings. The top chord, running between finialed tower caps, formed an ogival or pointed arch. Finally, McCullough disguised access openings for tower maintenance as semicircular arched windows that harmonized with the bracing panels.[13]

The second structure was the Río Choluteca Bridge near the village of Choluteca in Honduras. It was much like the Río Chiriquí Bridge structurally and architecturally. Differences were dictated by location and McCullough's penchant for variety. He decided to construct a two-span cable suspension structure with loaded backstays measuring 984 feet. The outermost towers were 660 feet apart with the central pier and tower halving the distance. This arrangement, McCullough argued, was the most efficient and economical plan for the site. He could have avoided a pier at mid-stream, but reasoned that then the structure would have

Río Choluteca Bridge, Inter-American Highway, Honduras, 1936.
(Federal Highway Administration)

cost more because of its structural steel and wire cable components. He also considered a structure with a 450-foot main span, but it required costly construction for two piers in the stream proper. The design he selected—the two-span cable suspension bridge—was the most prudent choice.[14]

Economics also affected McCullough's selection of a method to contain road deck deflection, or sagging, under live loads and his architectural pier treatment. On this and the bridge in Panama, McCullough avoided using costly live-load tower-to-deck stay cables or heavy steel deck plating. Instead, he employed deep stiffening trusses running the length of the structure and along both sides of the deck to maintain rigidity. The Río Choluteca's proximity to a Tufa stone quarry and inexpensive labor for masonry work persuaded McCullough to face the concrete piers and cable anchorages with the pinkish-gray rock. He believed the Tufa facing complemented the natural landscape.[15]

McCullough emphasized the Spanish Colonial style on the steel towers and handrails. Instead of the Gothic-arch portal opening seen on the Río Chiriquí Bridge, he chose a wide Romanesque arch for the Río Choluteca structure. Between this and a castellated top strut, he added an arcade of long, slender, semicircular arches. The box-steel suspension towers featured decorated inspection window cutouts. McCullough continued the Spanish Colonial theme in the steel handrails.[16]

The last of the three bridges was a 440-foot eyebar suspension structure over the Río Tamasalupa in Guatemala, near Anunción Mita. Its 240-foot main span was the shortest of the three, and it marked a departure in both structural and architectural design. The Río Tamasalupa presented obstacles similar to the other Central American streams. It was peaceful during the dry season, but swollen and swift during the rainy season. Compounding these variables was the river's unstable foundations that necessitated piling to secure bridge piers to bedrock. For crossings with this geological characteristic in Oregon (i.e., the Wilson River and Alsea Bay bridges), McCullough had chosen reinforced-concrete through tied arches because they did not require heavy piers to contain arch rib thrust.[17]

The problem on the Río Tamasalupa was similar and the self-anchoring bridge that McCullough designed was the suspension equivalent of the tied arch, only its action was the opposite. Where the deck of the tied arch functioned in tension to compress the arch ribs and contain their thrust, the self-anchoring suspension eyebars, acting in tension, transmitted their load to the stiffening trusses of the bridge's deck that acted in compression. The superstructure rested on lightly constructed piers.[18]

McCullough studied Mayan architecture and culture while designing the Río Tamasalupa Bridge. Its influences were apparent in his aesthetic treatment of the Guatemalan structure. "It seemed fitting to strike the Mayan note," McCullough explained, "even if faintly, in the treatment of the bridge." The "folded back" terminals or knee-brace-style ornaments between the top chord of the portal bracing and the 50-foot towers were, according to McCullough, "reminiscent of the conventionalized 'serpent's jaws'—a sculptural motif that persists with unusual frequency throughout the entire area of [Mayan] 'Old Empire' building." Additional bracing between the chevron-style top and intermediate struts continued the serpent-jawed theme, as did the handrail panels. These and other Mayan hieroglyphic designs appeared in relief on the bridge's concrete guardrails and integrated pylons.[19]

Construction of the three bridges and twelve other smaller structures created a substantial logistical problem. Only a coordinated effort between U.S. steel component suppliers, local transportation providers and laborers, and McCullough's staff engineers surmounted it. The construction of the Río Choluteca Bridge illustrates this.

Río Tamasalupa Bridge, Inter-American Highway, Guatemala, 1936.
(Federal Highway Administration)

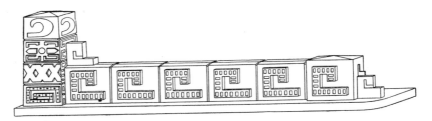

Río Tamasalupa Bridge: railing and pylon details.
(Federal Highway Administration)

The United States Steel Products Company (U.S. Steel) fabricated the box-steel components for the superstructure and the reinforcing steel for the piers. It shipped the material to Ampalas, Honduras, a port located on an island 16 miles from the mainland. Small vessels transported the shipments to the coast, where carts and trucks hauled them 25 miles to the construction site. There, dugout canoes ferried men and materials to the other side of the river. Meanwhile, local laborers quarried the Tufa, transported it to the river by ox cart, and carried half of it across the river for use on the piers and cable anchorages on that side. Then American and Honduran engineers assembled the bridge's steel superstructure, completing the construction process.[20]

As the three suspension spans and other bridges neared completion in January 1937, McCullough left Archibald in charge of the project and sailed for Portland, Oregon. Although Central America was "marvelous country," McCullough yearned for the temperate climate of Oregon. He docked in Portland on 31 January to meet a crowd of friends and family eager to hear about his sixteen months abroad. When McCullough arrived at Salem the next morning, 6 inches of snow greeted him, which grew nearly another 2 feet by evening. Trading tropical Central America for Oregon's unusually cold and snowy February foreshadowed a different kind of storm in McCullough's future as Oregon's bridge engineer.[21]

For the moment, McCullough had become something of a celebrity. He addressed various gatherings, where he presented travelogue slide shows of Central American landscapes and Mayan architecture, and he lectured on the Inter-American Highway and its effects on Central America. McCullough traveled to Marshfield, on Coos Bay, where dignitaries greeted him with a dinner and applause that acknowledged his preeminent role in constructing the five coast bridges. Little did he know at the time that they were the last he would ever design.[22]

⋈ ⋈

WHEN HE RETURNED TO OREGON, McCullough did not rejoin his bridge department. Instead, the state highway commission promoted him to full-time assistant state highway engineer. McCullough had received that title five years before, but it had carried no additional responsibilities or salary. Now, because of the state's growing highway system, the OSHC chose to harness McCullough's abilities as an administrator. He oversaw highway and bridge construction, maintenance budgets, and project planning. The OSHD's materials testing laboratory and the travel and information bureau reported to him, and he oversaw all personnel matters pertaining to employee evaluation and promotion. He was completely divorced from the bridge department's daily activities. Glenn Paxson became the new OSHD bridge engineer.[23]

McCullough resented being "kicked upstairs" after returning from Central America. Although the promotion had been given as a reward for years of successful bridge designing, he was frustrated because it offered little challenge. In addition, hostility existed between him and Robert Baldock, his superior, dating back to 1932, when he and Baldock had competed for the state highway engineer's post. Baldock, then a division engineer at Roseburg, had received the job. As something of a consolation prize, McCullough had been named assistant highway engineer, which did little to change his role as state bridge engineer.[24] Both were strong-willed men and their personalities clashed. Neither minced words and neither feared arguing a point. McCullough was stubborn, and when he thought he was right, he stood up for principle. Baldock and McCullough argued "round and round" over points of disagreement and went "toe to toe and eyeball to eyeball." McCullough never developed a strong commitment to his new position. His new-found autonomy, however, allowed him to resume researching and writing.[25]

During the next decade, McCullough wrote on numerous topics, including bridge engineering, but mostly on assorted personal interests involving the larger questions of highway management. He explored highway planning economics, highway revenue taxation structures, merit promotion systems for department personnel, and engineering law.[26]

Between 1938 and 1944, McCullough and several OSHD colleagues published five departmental technical bulletins on suspension bridges, based on the three structures built on the Inter-American Highway. They argued against the prevailing engineering theory, suggesting that suspension bridges were viable alternatives for crossings of less than 750 feet. Highway engineers avoided this type of bridge except to span extremely long distances, such as the Golden Gate or Oakland Bay crossings at San Francisco, because the mathematical analysis involved in their design was tedious. McCullough believed that suspension bridges were effective options, competing favorably with cantilever or trussed spans on costs.[27]

McCullough, Glenn Paxson, and Dexter Smith argued in favor of short-span bridges because more sophisticated mathematical calculations, using the Fourier series or sine series method of exact stress analysis, gave more accurate calculations than the traditional and less reliable elastic theory. The new method streamlined the designer's approach to determining a structure's specifications based on load requirements, roadway widths, and total length. McCullough, Paxson, and structural engineer Richard Rosencrans advanced this idea with experimental verification using scale models of suspension bridge designs.[28]

In terms of economy, the suspension bridge for short- and medium-length spans (up to 1,800 feet), was a worthy choice not only for lower material and labor costs, but also for its "revenue and benefit factors." McCullough and his OSHC colleagues believed that cantilever or truss spans were unequal to the "beauty of line and proportion inherent in the suspension bridge." Nor could they match the "graceful, sweeping cable curves" that presented "an esthetic [*sic*] silhouette." Thus the suspension bridge offered "aesthetic excellence" to travelers and local residents, potentially increasing traffic, gasoline tax revenues, and tourist spending. They suggested that bridge designers include in construction budgets a "rental value" of as much as 2 percent of the first cost of a structure for its aesthetic appeal, particularly for suspension bridges located on scenic routes.[29]

McCullough was involved with suspension bridges in two other areas. He was one of the leaders of a bi-state commission charged with investigating new bridge crossings over the Columbia River, between Oregon and Washington. In 1944, the Oregon State Highway Department and the Washington Department of Highways issued a report entitled, *Trans-Columbia River Interstate Bridge Studies*, in which they recommended construction of suspension bridges at several crossings between Umatilla and Astoria. Interestingly, the commission contemplated a suspension bridge at the river's mouth, but "the advent of war, however, and the attendant possibility of danger from bombing raids" caused them to look inland, 13 miles east of Astoria, at the Jim Crow Crossing, for a bridge site. And there, the commission proposed a steel cantilever rather than a suspension span.[30]

McCullough and Rosencrans were also members of the commission that investigated phenomena associated with the 1940 failure of Leon Moisseiff's Tacoma Narrows Bridge. This culminated in a 1950 BPR-sponsored report entitled *The Mathematical Theory of Vibration in Suspension Bridges*.[31]

❧❧

McCULLOUGH DEVOTED HIMSELF to activities outside the highway department to expend his stores of energy. For him, work was relaxation, whether at home, at church, or for the community. In the 1930s, McCullough commissioned noted Salem architect Clarence Smith to design a comfortable Tudor-style home at 285 West Lefelle Street, in south Salem's Fairmount Addition. In his den, McCullough even tinkered with writing fiction modeled after the serialized detective stories found in popular magazines of the 1930s and 1940s. In his stories, an engineer was the main character, someone who solved difficult problems through his knowledge of structural engineering. None were published.[32]

He also became a regular churchgoer. McCullough's childhood had been filled with religious activities, so much so that when he left home he tried to "get as far away from it as possible so [he] became an engineer." In Salem, he and the local Episcopal rector, George Swift, became close friends. By 1940, McCullough, his wife, and son were confirmed as Episcopalians at St. Paul's Church, in Salem. Shortly thereafter, the congregation elected McCullough to the parish vestry, or governing body, where he served for several years.[33]

McCullough had been a member of the Salem Chamber of Commerce's Planning Committee since 1925. He had worked with city engineer Ray A. Furrow in designing twelve reinforced-concrete bridges on Salem's arterial streets in the late 1920s and early 1930s. Many of these girder spans included McCullough's signature details, including arched curtain walls, bush-hammered inset panels, and precast ornamental railing panels.[34]

McCullough's involvement in civic activities was somewhat unusual for a state employee. Many Salem residents, especially businessmen, resented anyone on the state payroll. They treated these public servants as second-class citizens, believing them to be overpaid and underworked. But McCullough pursued his civic activities as he did anything else— with great energy. He and his wife also became close friends with several downtown shopkeepers, and he joined the local Rotary Club chapter.[35]

Years of sixteen-hour workdays, endless packs of cigarettes, and the embattled office atmosphere took their toll on McCullough's health. He also feared for his son's safety when the young man enlisted in the Navy shortly after the Japanese attacked Pearl Harbor, but wrote to encourage the young man: "[R]emember, that you and all of the rest are paying your debt to the Country which has given you the opportunity to live and laugh and learn and become individuals, and if you and all of the rest pay your debt cheerfully, with obedience in spirit as well as deed, the lazy tranquillity of the peace-time days will come back again." McCullough suffered a heart attack in early 1942, and his life changed abruptly. Doctors ordered him to slow down and quit smoking. McCullough temporarily followed the prescribed regimen. His wife and son worried about his condition, but he remained cynical about it. "I am going to live to be 95 years old and will raise so much hell," McCullough boasted, "that you will be sick and tired of me long before I pass out of the picture." He resented the tight reins placed on his active lifestyle. "The hell with this," he announced. "If I have to live like this the rest of my life, I don't want to live. I want to enjoy life." Accordingly, he resumed his old habits.[36]

During the mid-1940s, nothing rewarded McCullough with more pleasure and intellectual stimulation than two dissimilar projects: his joint authorship with his son, John, of a two-volume text entitled *The Engineer at Law: A Resumé of Modern Engineering Jurisprudence*, and his leadership of Salem's new Long Range Planning Commission.

McCullough had maintained his interest in law since his years with the Iowa State Highway Commission. Repeated involvement in cases ranging from patent litigation to right-of-way acquisition strengthened his belief in the need for a standard work for practitioners of both professions. In fifteen chapters comprising 800 pages, he examined all areas where engineering and law converged, i.e., contracts, engineering specifications, real property law, torts, patents, and testimony. Oregon Supreme Court Justice James T. Brand wrote in the foreword, "The book presents a bird's-eye-view of the broad field of law," for the engineer and lay person who "desires a sympathetic understanding of our legal institutions." Brand believed the book was also a "power house of applied information geared directly to engineering problems." It was "a guide" for the engineer "away from lawsuits, not through them."[37]

As World War II came to an end, Salem's Chamber of Commerce focused on the impending postwar growth and the avoidance of haphazard development. Its population had grown from 20,000 in the 1920s to nearly 40,000 by 1945. The city needed an "ordered and scientific" plan, so the chamber asked McCullough to chair its new Long Range Planning Commission. McCullough and other commission members gathered statistical data on traffic, enlisted highway department designers for proposed road realignments, and studied projected population demographics. They assembled a wealth of material on city zoning regulations, public building construction, and parks systems. They surveyed Salem's land use regulations and investigated zoning proposals for the greater metropolitan area beyond the city boundaries.[38]

The commission promoted industrial park development, off-street automobile parking in business districts, and the designation of arterial streets for efficient downtown traffic flow. It also sought to alleviate Salem's nightmarish ring of rail lines surrounding the business district, which created traffic hazards and retarded speedy responses to fires and medical emergencies. It sought improved transportation networks, including better airport facilities and an east-side bypass of the city for the Pacific Highway. Finally, the commission believed that government's presence in Salem needed to grow in an orderly fashion. It urged the coordination of city, county, and state planning for new government buildings to preserve the city's aesthetic qualities. McCullough and his associates produced a document with vision.[39]

Meanwhile, Thomas H. MacDonald contacted McCullough and invited him to return to Central America in May 1946 to help the BPR again in its construction of the Inter-American Highway. More bridges were needed in El Salvador and Honduras. It was an opportunity McCullough regarded with great enthusiasm. The Iowa State College Press, in cooperation with the OSHD, had just released his book, which was receiving favorable reviews from legal and engineering critics alike. The planning commission's progress report was coming together nicely. Returning to Central America offered him the chance to resume his passion, building bridges.[40]

On Sunday, 5 May 1946, McCullough collapsed at his Lefelle Street home after an afternoon of gardening. He suffered a massive cerebral hemorrhage that paralyzed his right side. He was rushed to Salem General Hospital, where he died at three o'clock the next morning. He never regained consciousness. Newspapers throughout the state reported his death as a tragic event that cut short a brilliant career.[41]

McCullough's funeral took place two days later, on the afternoon of Wednesday, 8 May, at St. Paul's Episcopal Church. State highway commission offices closed that day, allowing personnel to attend the service. The church overflowed with friends and colleagues, and hundreds stood outside. Highway department officials serving as pallbearers included Robert Baldock, Joseph Devers, and William Reeves. The honorary pallbearers included representatives of the judiciary, newspapers, and commissioners and employees of the state highway department. The

Reverend George Swift read the burial office and paid tribute to McCullough as a "man of brilliant intellect, generosity, fellowship, democracy and deep religious convictions."[42]

Subsequent editorials echoed Swift's assessment. Salem's *Oregon Statesman* eulogized McCullough as "a gallant soul who lived life to the full" and continued, "He gave much from his great mind and overflowing heart, and from life he derived rich satisfaction in achievement...." The Salem *Capital Journal* stated, "In the spirit of democracy which he created about him he was 'Mac' to all with whom he came in contact, even in the closely bound circle of internationally famed structural engineers in which he moved on equal footing with the most famous." William Tugman, a close personal friend, wrote in the Eugene *Register-Guard* that McCullough was more than a famous bridge designer. He was "a man in whom intellectual brilliance was blended with those rare qualities of character which cause a man to be beloved by associates and friends." Finally, Don Upjohn, a columnist for the *Capital Journal*, summarized McCullough in this way: "He's one of the very few who cannot be replaced."[43]

Two events in 1947 formed happy epilogues to McCullough's productive life. In January, the Salem Chamber of Commerce released his brainchild document, *A Long Range Plan for Salem, Oregon: First Annual Progress Report*. The 84-page volume was dedicated to McCullough's tireless contributions to planning Salem's future. Many of the commission's suggestions came to fruition as the city grew to over 100,000 inhabitants. Two of the four rail lines encircling the central business district were removed and arterial traffic routes were improved. The Pacific Highway moved east of the city in the early 1950s and became Interstate 5.

On Wednesday, 27 August 1947, T. H. Banfield, OSHC chairman, and representatives of Oregon State College unveiled a plaque on the mile-long Coos Bay Bridge. It dedicated "this great structure" as the "Conde B. McCullough Memorial Bridge," to a man "whose genius and inspiration are manifest" in its design and that of many others in Oregon. In his 1995 volume *Engineers of Dreams: Great Bridge Builders and the Spanning of America*, Henry Petroski wrote that the "Conde B. McCullough Memorial Bridge thus joined the exclusive group that includes the Eads Bridge at St. Louis and the Roebling Bridge at Cincinnati in being named for its engineer."[44]

❧ 8 ❧

Conclusion

C ONDE B. MCCULLOUGH HAD AN EVENTFUL and rewarding career in bridge engineering and highway building and he became one of the great American civil engineers of the twentieth century. His tutelage under Anson Marston, who harnessed his students' innate abilities to "fit them to be engineers," came at the time when modern American highway bridge building was at its beginning. Important technical advances in reinforced-concrete construction in Europe and the United States appeared just as McCullough began his professional training.

The engineering skills McCullough learned at ISC prepared him for an apprenticeship with Marsh Engineering when James Marsh was designing, building, and patenting his rainbow arch. This experience reinforced McCullough's growing commitment to the advantages of modern concrete construction. He also learned firsthand how private bridge engineers incorrectly claimed that their designs were genuine engineering advances rather that new applications of old technology.

Seven years as bridge engineer and assistant highway engineer with the ISHC provided McCullough with the opportunity to go beyond the short and simple research projects of his college days. Instead, he conducted new investigations on a grand scale, hoping to increase the knowledge of reinforced-concrete bridges for the growing network of state highway departments. Working with like-minded engineers such as E. Foster Kelley, John H. Ames, and Thomas H. MacDonald, McCullough argued that standardized, well-designed bridges were necessary for Iowa's road system and that an efficient highway agency was required to oversee their construction and maintenance.

Iowa's involvement in the lawsuit *Luten v. Marsh Engineering* provided McCullough with the opportunity to research the history of worldwide reinforced-concrete construction. He broadened his knowledge of the field and applied it throughout his career. His participation in the lawsuit also introduced him to the field of engineering law, which he enthusiastically pursued during his years in Oregon.

McCullough's tenure as professor at Oregon Agricultural College afforded him a brief opportunity to teach undergraduates about engineering in the same way as he had learned it from Marston. He embraced and promoted the civil engineering philosophy that Estévan Fuertes had passed on to Anson Marston at Cornell University—that engineers were not simply technicians but specialists who needed to be broadly educated. McCullough believed that all fields of knowledge overlap, and that to be an effective engineer one needed an understanding of other fields of inquiry.

The years at OAC also provided McCullough the opportunity to observe the state's creation of a progressive highway organization. The Oregon State Highway Commission sought to provide the public with efficient and economical bridge and road structures that were delicately balanced with aesthetic considerations. Federal assistance to state highway organizations, begun in 1916, arrived just as Oregon's citizens had mandated an intensive state highway improvement program and the nation's first gasoline tax. When offered the opportunity to become the OSHC bridge engineer, McCullough accepted without hesitation.

McCullough was energized by the great challenges he saw in Oregon bridge engineering. He met each challenge with the will to create efficient bridges that would provide long service at little or no additional cost to the public. Where possible and when the location justified it, he designed bridges that afforded elegant architectural presentations. All three components, he believed, went hand-in-hand, and an improved structural design would yield a bridge that would be, in the long run, less expensive than cheaper alternatives. When dealing with reinforced-concrete and steel construction, he saw it as his mission to create structures that were not intrusive to their natural settings. Engineers, he believed, had too frequently "cluttered up the landscape," and the simple elegance of his bridges provided an architectural antidote.

McCullough's preferred design form was the arch, which he used throughout the 1920s and 1930s because of its practicality and grace. Like the searching artist, he was never satisfied with his creations, always researching and looking for ways to achieve greater utility and architectural refinement. McCullough's experimentation with gunite-covered steel, the Freyssinet precompression technique, hinge types, and tied-arch construction, all attested to his preeminence as a highly skilled bridge designer. Bridges at Oregon City and the numerous structures along the Oregon Coast Highway confirm this claim.

Volumes of writings in professional engineering journals, book chapters, state and federal technical bulletins, and textbooks illustrate McCullough's stature as a respected member of his profession. In doing so, he lived up to Anson Marston's essential qualities of an engineer. McCullough combined his training in the theoretical fundamentals of engineering with skills that he learned as a practitioner, and he gave fellow engineers the benefit of his experiences through his writings. Professional recognition by his peers provided yet another measure of his achievement. The prestigious American Institute of Steel Construction named his McLoughlin Bridge the most meritorious structure in its class in 1933, and OAC awarded him an honorary doctorate in 1934 for his stature as an "international authority on bridge design."

McCullough's plans for the five Public Works Administration-sponsored coast highway bridges in the mid-1930s illustrated his maturity as an engineer at the pinnacle of his career. He moved well beyond treating the reinforced-concrete arch as a variation on masonry bridges and incorporated it as a form fundamental to the plastic nature of this building material. These bridges represent some of the most architecturally significant reinforced-concrete structures in the nation. They marked the heyday of reinforced-concrete arch construction in American highway building, which ended with the beginning of World War II.

Ironically, Eugène Freyssinet's 1920s innovations in reinforced-concrete arch construction led to advances in bridge design that made the prestressed reinforced-concrete girder span, either pre-tensioned or post-tensioned, the preferred bridge type during the postwar American highway building boom. The arch became reserved only for special structures, never again to be seen on the American landscape with such regularity.[1]

Oregon bridge building was never the same after McCullough's 1937 promotion to assistant state highway engineer. Under Glenn Paxson, the bridge department created unimaginative structures with almost no regard for aesthetics. Paxson's business-as-usual comments about the Coos Bay Bridge's architectural qualities reflected as much his own viewpoint as it did a change of priorities in American bridge design philosophy.

Meanwhile, McCullough reluctantly moved away from designing Oregon's bridges to administration. He immersed himself in a study of short-span suspension structures because he hoped that other state bridge engineers would take seriously his belief that they could be as much a part of modern highway design as the larger, grander bridge types. Through deliberate isolation from his superior, Robert Baldock, McCullough continued his productivity as a researcher. His long tenure in state-sponsored highway building programs in Iowa and Oregon gave him an enlightened perspective on management issues. He demonstrated, moreover, how public agencies that were expected to operate efficiently and economically could produce public works projects of grace and beauty. Conde B. McCullough was one of the few twentieth-century bridge engineers who left such an enduring mark on the American landscape.

Epilogue

⚜   ⚜

O F CONDE B. McCULLOUGH'S MAJOR WORKS in Oregon, only a handful have been lost. Some, however, have been modified with new parapet walls or additions that detract greatly from their original composition. The loss of one structure in particular, the Alsea Bay Bridge, proved the impetus for Oregonians and the Oregon Department of Transportation (ODOT) to seek methods to preserve other monumental McCullough spans.

Sadly, the hostile salt air environment at the oceanside caused extensive damage to McCullough's bridge at Waldport during the fifty-five years that it carried vehicles across Alsea Bay. In 1972, active corrosion of the steel reinforcing in the pier foundations prompted ODOT to begin a cathodic protection program to slow the deterioration. The technology known at the time, however, was not enough to save the bridge. By the early 1980s, the bridge was in desperate shape. Spalling on the deck's underside had become so advanced that falling concrete posed a threat to boaters. Soon, load restrictions were posted. After a long project development process—including the release of an environmental impact statement, public participation, and special studies—ODOT, in the mid-1980s, decided against rehabilitating the old structure and chose instead to replace it. The national engineering consulting firm of HNTB (Howard, Needles, Tammen, and Bergendoff) prepared several conceptual designs. A citizens' advisory committee assisted ODOT in the selection process. After evaluating many options, including cable-stayed spans, girder spans, deck arches, and through arches, ODOT, with the blessing of Waldport's townsfolk, replaced the decaying bridge with a 2,910-foot structure, including a 350-foot steel through-arch resting on Y-shaped piers. Groundbreaking on the $42.6 million bridge took place in 1988.[1]

Innovations incorporated into the new bridge's design prevent or at least delay the disastrous effects of chloride ion penetration (from salt,

NaCl) that had affected the old bridge. All reinforcing steel was coated with epoxy to make it corrosion resistant. The road deck's wearing surface was made of latex concrete; and the outer covering of the piers was a layer of concrete at least four inches thick, rather than the normal one-inch layer, to keep reinforcing steel well away from the surface, where moisture and corrosive salt air might penetrate. The special procedures on the new Alsea Bay Bridge, which was completed in the fall of 1991, give it a seventy-five- to one-hundred-year life expectancy. Demolition of the old bridge climaxed on 1 October 1991 with the dynamiting of two of the three tied-arch spans.[2]

In 1990, ODOT contracted with the National Park Service's Historic American Engineering Record (HAER) to study twenty-five historic highway bridges throughout the state. The agency wanted to develop a better understanding of these structures and to gather material useful in future mitigation for possible construction-related alterations to their original designs. Historians, architects, and photographers documented the bridges with written reports, measured and interpretive drawings, and large-format black-and-white photography, which became part of HAER's permanent collection at the Library of Congress.

Many of the bridges studied were structures along the Oregon Coast Highway designed by McCullough. Seven drawings focused on these spans. One compared the major McCullough bridges on the Oregon Coast Highway. Two focused on the Alsea Bay Bridge and its reinforced-concrete tied-arch center spans. Four additional drawings illustrated the Freyssinet precompression technique that McCullough employed on the Isaac Lee Patterson Bridge, also known as the Rogue River Bridge.

During the course of construction on the new Alsea Bay Bridge, workers also erected the Alsea Bay Interpretive Center, a wood and steel-roofed structure designed by the ODOT bridge section, at the southwest corner of the new span. A memorandum of agreement between the Federal Highway Administration, the Advisory Council on Historic Preservation, and the Oregon State Historic Preservation Office, stipulated the construction of the $788,000 building as part of mitigation for destruction of the old bridge—a National Register-eligible structure. The Interpretive Center, under the jurisdiction of the Oregon Parks and Recreation Department, was dedicated and opened to the public on 2 November 1991. The seven HAER drawings became a focal point in the center's exhibit hall.

Portions of the old bridge were salvaged and reused on and around the new span. Two of the original pylons, cedar spires, and some railing were incorporated into the Alsea Bay Bridge North Wayside, located near the north end of the new bridge, on the original highway alignment to the old structure. The southern pylons were reused at the south end of the new bridge, and near the Interpretive Center. Portions of railing were installed in the exhibit space.[3]

In the 1980s, ODOT recognized that a significant number of its historic highway structures—along the Oregon Coast Highway, the Historic Columbia River Highway, and in the Portland metropolitan area—were deteriorating in ways conventional methods could not control. The spectacular beauty of McCullough's arch spans on the Oregon Coast Highway and the fear that they would be lost to corrosion damage compelled ODOT to make a fundamental change in its approach to historic bridges. The department undertook a comprehensive inventory of its older bridges, identifying those which had true historic features and should be preserved. Of over a thousand extant bridges in Oregon constructed before 1941, the study identified one hundred forty-five as eligible for listing in the National Register of Historic Places and fifty-three as potentially eligible. As a result, ODOT began to restore, rather than rehabilitate or replace, these bridges.

The department hoped to prevent a repeat of the painful episode with the Alsea Bay Bridge. A new engineering unit within the ODOT Bridge Section (the legacy of McCullough's "Bridge Department") performed a thorough evaluation of the eighteen arch bridges on the Oregon Coast Highway, because they were most at risk. ODOT then took a dramatic step forward in historic bridge preservation, developing and employing methods to halt the ocean-induced damage to the Oregon Coast Highway bridges, restore them to their original condition, and preserve them from future corrosion damage. The department demonstrated practicable solutions, often accomplished for less than the cost of replacing the structures.

The bridge section's program employs practical methods for shotcrete and pumped concrete repairs, precasting of replacement components, composite strengthening, and cathodic protection of structures. The goal is to preserve both the intended function and the original construction of Oregon bridges formally identified as significant historic resources. The policy results from the demands of the citizens of Oregon that the department do more than just meet legislative requirements or design standards.

The cathodic protection process begins when deteriorated concrete is removed and replaced. Patching materials must closely approximate the strength and conductivity of the original concrete or they will accelerate the existing corrosion and cause the patched area to spall off within a few years.Shotcrete, pneumatically applied concrete, appears to work very well in Oregon. Shotcrete is very easy to hand tool to restore architectural detail lost with the removal or spalling of original concrete. Placing forms and pumping concrete also works well for less detailed areas. ODOT adds salt to the concrete so that its conductivity will be similar to that of the underlying concrete, which has absorbed salt from the sea air. Reinforcing steel within concrete will corrode and ultimately fail, especially if chlorides (for example, salt) are allowed to infiltrate the concrete. Placing another, chemically more active, metal at the surface of the concrete forces that metal to corrode instead of the reinforcing steel. The steel that is protected functions as the cathode and the metal that corrodes sacrificially functions as the anode. Connecting a direct current power supply between the two improves the protection. For the Oregon Coast Highway bridges, ODOT most often employs arc-sprayed zinc as the anode in its cathodic protection program because it is applied much like spray paint, faithfully preserving intricate architectural surface details.

When this book went to press, ODOT had completed work on the bridges at Yaquina Bay, Depoe Bay, Cape Creek, and Big Creek. Projects were under development for the bridges at Rocky Creek, Cummins Creek, and the Rogue River. Only eleven McCullough-designed bridges remained to be preserved along the Oregon Coast Highway and the department planned to restore other historic bridges from throughout the state by 2010. Historically accurate reproduction rail replacement had been completed or was being developed for additional bridges.[4]

In the 1990s, ODOT decided to replace McCullough's 1926 Crooked River (High) Bridge, a steel deck arch near Terrebonne, on U.S. 97. The two-lane structure was narrow by current standards. David Goodyear Engineering Services and ODOT's bridge section designed the new bridge, a four-lane, 535-foot-long reinforced-concrete deck arch. Kiewit Construction began erecting the $18.3-million bridge in late 1997 and it opened nearly three years later. McCullough's original bridge remains in place, and functions as an extension of the adjacent Peter Skene Ogden State Scenic Viewpoint, offering pedestrians and bicyclists breathtaking views of the Crooked River Gorge along with the new bridge to the east, and Ralph Modjeski's 1911 steel deck arch railroad bridge to the west.

❧❧

SINCE THE MID-1980S, WHEN ODOT completed its comprehensive survey of Oregon's historic highway bridges, McCullough and his bridges have begun receiving much overdue attention from historians, engineers, and architects. In 1985, the Oregon Historical Society and ODOT published *Historic Highway Bridges of Oregon*. The volume, which appeared in a second edition in 1989, is an outgrowth of the survey and draws attention to the best of Oregon's bridges, many of which are structures designed by McCullough. In 1993, Eric N. DeLony published *Landmark American Bridges*, a volume showcasing the best of American bridge construction. He saw McCullough's work as "a remarkable outpouring of creativity and skill matched by few state highway departments in the country." He believed that McCullough's Oregon Coast Highway structures are "some of the best and most innovative concrete and steel bridge designs in the world."[5]

In 1998, the Oregon State University College of Engineering inducted McCullough into its Engineering Hall of Fame. Membership is granted to those OSU alums and others "who have made sustained and meritorious engineering and/or managerial contributions throughout their careers."

In 1999, in honor of its one hundred twenty-fifth anniversary, the periodical *ENR*, once known as *Engineering News-Record*, published a list of the top people who had made outstanding contributions to the construction industry since 1874. "Their efforts," *ENR* believed, "helped shape this nation and the world . . . by developing new analytical tools, equipment, engineering or architectural design." Ten bridge engineers made the list, and among them was C. B. McCullough. "These leading designers dared to span great lengths," wrote *ENR*, "with the most elegant, constructible and economical solutions possible." The periodical cited McCullough's use of the reinforced-concrete tied arch as his most innovative contribution. He was listed among such engineering greats as Othmar H. Ammann, James Eads, Robert Maillart, and David B. Steinman.[6]

McCullough's Mosier Creek Bridge and Dry Canyon Creek Bridge on the Columbia River Highway are included in a section of the route stretching from Troutdale to The Dalles known as the "Historic Columbia River Highway" (HCRH). These bridges and the Mosier Twin Tunnels, designed by his department, contributed greatly to the HCRH's 1983 listing in the National Register of Historic Places. A year later, the American Society of Civil Engineers designated the HCRH a "National Historic Civil Engineering Landmark." In 2000, much of the HCRH, including McCullough's two arches, was designated a National Historic Landmark, in part for its contributions to early modern road building in the United States and its sensitivity to the natural landscape.

Interestingly, with all the attention given to McCullough and his long list of monumental bridges in Oregon, none of his spans is listed individually, or as part of a multiple-property submission, in the National Register of Historic Places. It would be very appropriate for fifty of his best bridges to be listed in a multiple-property nomination, both for their significance as examples of modern American highway bridge engineering and as examples of McCullough's work. More importantly, five of McCullough's structures—the Isaac Lee Patterson Bridge and the four remaining spans from the Oregon Coast Highway bridge project—should receive National Historic Landmark status. They are truly worthy of this designation because they are nationally significant examples of early-twentieth-century highway bridge engineering and the most notable examples of McCullough's engineering legacy. National Historic Landmarks serve as guides in comprehending important trends and patterns in American history, offering the opportunity to better understand the conditions that shaped the country. Conde B. McCullough's nearly two decades of innovative bridge design in Oregon surely merit this designation.

Appendix: Major McCullough Bridges

Arches

YEAR	ODOT NO.	NAME AND DESCRIPTION
1920	00332A	**Rogue River (Rock Point) Bridge**, Jackson County. One 113-foot reinforced-concrete deck arch; total length 505 feet.
1920	00409	**Sucker Creek (Oswego Creek) Bridge**, Clackamas County. One 130-foot reinforced-concrete deck arch; total length 330 feet.
1920	00498	**Mosier Creek Bridge**, Wasco County. One 110-foot reinforced-concrete deck arch; total length 182 feet.
1921	00524	**Dry Canyon Creek Bridge**, Wasco County. One 75-foot reinforced-concrete deck arch; total length 101 feet.
1922	00490A	**South Umpqua River (Myrtle Creek) Bridge**, Douglas County. Three 130-foot reinforced-concrete deck arches; total length 547 feet.
1922	00357	**Willamette River (Oregon City) Bridge**, Clackamas County. One 360-foot steel half-through arch; total length 745 feet.
1924	00839	**North Umpqua River (Winchester) Bridge (Robert A. Booth Bridge)**, Douglas County. Seven 112-foot reinforced-concrete deck arches; total length 884 feet.
1924	00626	**Grand Ronde River Bridge (Perry Overcrossing)**, Union County. One 134-foot reinforced-concrete deck arch; total length 312 feet.

YEAR	ODOT NO.	NAME AND DESCRIPTION

1925   00624A   **Umatilla River (Umatilla) Bridge (William Duby Bridge)**, Umatilla County. Three 110-foot reinforced-concrete deck arches; total length 439 feet. Removed during a widening project in 1951.

1925   01095   **Fifteenmile Creek (Adkisson) Bridge**, Wasco County. One 120-foot reinforced-concrete deck arch; total length 148 feet.

1926   00600   **Crooked River (High) Bridge**, Jefferson County. One 330-foot steel deck arch; total length 464 feet.

1927   00576   **Rogue River (Gold Hill) Bridge**, Jackson County. One 143-foot reinforced-concrete barrel arch; total length 443 feet.

1927   02459   **Depoe Bay Bridge**, Lincoln County. One 150-foot reinforced-concrete deck arch; total length 312 feet. Widened 1940.

1927   01089   **Rocky Creek Bridge (Ben F. Jones Bridge)**, Lincoln County. One 160-foot reinforced-concrete deck arch; total length 360 feet.

1928   01319   **Soapstone Creek Bridge**, Clatsop County. One 108-foot reinforced-concrete deck arch; total length 152 feet.

1930   00986   **Klamath River (Keno) Bridge**, Klamath County. Three 100-foot reinforced-concrete through arches; total length 370 feet. Destroyed March 1985.

1931   01499   **Wilson River Bridge**, Tillamook County. One 120-foot reinforced-concrete through tied arch; total length 180 feet.

1931   01181   **Ten Mile Creek Bridge**, Lane County. One 120-foot reinforced-concrete through tied arch; total length 180 feet.

1931   01180   **Big Creek Bridge**, Lane County. One 120-foot reinforced-concrete through tied arch; total length 180 feet.

YEAR	ODOT NO.	NAME AND DESCRIPTION
1931	01418	**Rogue River Bridge (Caveman Bridge)**, Josephine County. Three 150-foot reinforced-concrete half-through arches; total length 550 feet.
1931	01182	**Cummins Creek Bridge**, Lane County. One 115-foot reinforced-concrete deck arch with reinforced-concrete deck girder approach spans; total length 185 feet.
1932	01172	**Rogue River (Gold Beach) Bridge (Isaac Lee Patterson Bridge)**, Curry County. Seven 230-foot reinforced-concrete deck arches; total length 1,898 feet.
1932	01600	**Hood River (Tucker) Bridge**. One 100-foot reinforced-concrete deck arch; total length 188 feet.
1932	01113	**Cape Creek Bridge**, Lane County. One 220-foot reinforced-concrete deck arch, 399 feet of reinforced-concrete deck girder spans on concrete columns; total length 619 feet.
1933	01582	**Santiam River (Jefferson) Bridge (Jacob Conser Bridge)**, Marion and Linn counties. Three 220-foot reinforced-concrete through arches; total length 780 feet.
1933	01617	**Clackamas River (McLoughlin) Bridge**, Clackamas County. Two 140-foot and one 240-foot steel through tied arches and four 50-foot reinforced-concrete deck girder spans; total length 720 feet..
1934	01923	**South Umpqua River (Winston) Bridge**, Douglas County. Three 180-foot steel through tied arches; total length 540 feet.
1936	01822	**Umpqua River (Reedsport) Bridge**, Douglas County. One 430-foot steel through truss tied arch swing span, four 154-foot reinforced-concrete through tied arches; total length 2,206 feet.
1936	01821	**Siuslaw River (Florence) Bridge**, Lane County. Oone 140-foot double-leaf bascule steel draw span, two 154-foot reinforced-concrete through tied arches; total length 1,568 feet.

YEAR	ODOT NO.	NAME AND DESCRIPTION

1936   01746   **Alsea Bay (Waldport) Bridge**, Lincoln County. One 210-foot and two 154-foot through tied arches, six 150-foot deck arches, all reinforced-concrete; total length 3,011 feet. Destroyed and replaced, 1991.

1936   01820   **Yaquina Bay (Newport) Bridge**, Lincoln County. One 600-foot steel through arch, two 350-foot steel deck arches, five 265-foot reinforced-concrete deck arches; total length 3,223 feet.

1936   02063   **Eagle Creek Bridge**, Multnomah County. Two 142-foot and one 182-foot steel through tied arches; total length 466 feet. Dismantled in 1969. Piers reused for reinforced-concrete deck girder structure at the same site. One arch re-erected in 1970 as the Clackamas River (Barton) Bridge, in Clackamas County.

Trusses

YEAR	ODOT NO.	NAME AND DESCRIPTION

1921   00441   **North Yamhill River Bridge**, Yamhill County. One 80-foot steel Warren deck truss, seven 40-foot reinforced-concrete deck girder spans; total length 360 feet.

1922   00799   **Grand Ronde River (Old Rhinehart) Bridge**, Union County. One 142-foot steel Warren deck truss, 178 feet of reinforced-concrete deck girder spans; total length 320 feet.

1925   01025   **Willamette River (Albany) Bridge**, Linn and Benton counties. Four 200-foot steel Parker through trusses, 290 feet of reinforced-concrete deck girder approach spans; total length 1,090 feet.

1925   00603   **Calapooya Creek (Oakland) Bridge**, Douglas County. One 100-foot steel Warren deck truss, nine reinforced-concrete deck girder approach spans; total length 473 feet.

YEAR	ODOT NO.	NAME AND DESCRIPTION
1928	01356	**Santiam River (Cascadia Park) Bridge**, Linn County. One 120-foot timber and steel Howe deck truss.
1929	01223	**Willamette River (Springfield) Bridge**, Lane County. One 550-foot steel continuous through truss with reinforced-concrete deck girder approach spans; total length 1,090 feet.
1929	00966	**Deschutes River (Maupin) Bridge**, Wasco County. One 200-foot steel Warren deck truss, thirteen reinforced-concrete deck girder approach spans; total length 826 feet.
1929	01318	**Umpqua River (Scottsburg) Bridge**, Douglas County. Three-span, 643-foot continuous steel through truss.
1931	01614	**Elk Creek (First Crossing) Bridge**, Douglas County. One 140-foot steel Warren deck truss.
1931	01465	**Elk Creek (Third Crossing) Bridge**, Douglas County. One 140-foot steel Warren deck truss.
1931	01406	**Elk Creek (Fourth Crossing) Bridge**, Douglas County. One 100-foot steel Warren deck truss.
1936	01823	**Coos Bay (McCullough Memorial) Bridge**, Coos County. One 793-foot and two 457-foot steel cantilever truss spans, thirteen 265-foot reinforced-concrete deck arches; total length 5,305 feet.

Movable Truss Spans

YEAR	ODOT NO.	NAME AND DESCRIPTION
1921	00330	**Old Young's Bay Bridge**, Clatsop County. Two 75-foot steel central bascule spans, fifty-eight pile trestle secondary spans and ten timber stringer spans of 1,616 feet; total length 1,766 feet.

YEAR	ODOT NO.	NAME AND DESCRIPTION
1922	00598	**Coquille River Bridge**, Coos County. One 235-foot steel Parker through truss swing span. Dismantled, 1990s.
1924	00711	**Lewis and Clark River Bridge**, Clatsop County. One 112-foot steel central bascule span, forty-eight pile trestle and stringer spans of 716 feet; total length 828 feet.

Reinforced-concrete Deck Girder Spans

YEAR	ODOT NO.	NAME AND DESCRIPTION
1920	00308	**Fifteenmile Creek (Seufert) Viaduct**, Wasco County. One 22-foot span, five 40-foot spans; total length 222 feet.
1920	00464	**Mill Creek (West Sixth Street) Bridge**, Wasco County. One 124-foot span.
1931	01601	**Elk Creek (Second Crossing) Bridge**, Douglas County. Six spans totaling 290 feet.

Suspension Bridges, Inter-American Highway

YEAR	NAME AND DESCRIPTION
1936	**Río Chiriquí Bridge**, Panama. Single-span 400-foot cable suspension bridge with unloaded backstays; total length, including short enclosed girder approaches, 655 feet.
1936	**Río Choluteca Bridge**, Honduras. Two 330-foot cable suspension spans with loaded backstays; total length 984 feet.
1936	**Río Tamasalupa Bridge**, Guatemala. Single-span 240-foot eyebar suspension bridge, with 100-foot side spans; total length 440 feet. Rumored destroyed during civil unrest.

❧ *Notes* ❧

Chapter 1

[1]C. B. McCullough to J. E. Mackie, 7 September 1937, Office of General Files, ODOT (quote).

[2]David Plowden, *Bridges: The Spans of North America* (New York: W. W. Norton, 1974), 66, 67 (quote).

[3]Emory L. Kemp, *West Virginia's Historic Bridges* (Charleston: West Virginia Department of Culture and History, West Virginia Department of Highways, and Federal Highway Administration, 1984), 120.

[4]David P. Billington, "History and Esthetics in Concrete Arch Bridges," *Journal of the Structural Division of the American Society of Civil Engineers* 103, no. ST11 (November 1977): 2138-40.

[5]For a discussion of the engineers' role during the Progressive Era, see John C. Burnham, "Essay," in *Progressivism*, by John D. Buenker, John C. Burnham, and Robert M. Crunden (Cambridge, MA: Schenkman Publishing Co., 12-20. Other works to consult on the idea of efficiency and scientific management of society are Samuel Haber, *Efficiency and Uplift, Scientific Management in the Progressive Era, 1890-1920* (Chicago: University of Chicago Press, 1964); Samuel P. Hays, *Conservation and Gospel of Efficiency: The Progressive Conservation Movement, 1890-1920* (Cambridge, MA: Harvard University Press, 1959); and Edwin T. Layton, Jr., *The Revolt of the Engineers: Social Responsibility and the American Engineering Profession* (Cleveland: The Press of Case Western Reserve University, 1971), especially chapters 5 and 6. See also Robert H. Wiebe, *The Search for Order, 1877-1920* (New York: Hill and Wang, 1967).

[6]Friedrich Bleich, Conde B. McCullough, Richard Rosencrans, and George S. Vincent, *The Mathematical Theory of Vibration in Suspension Bridges,* Bureau of Public Roads, Department of Commerce, (Washington DC: Government Printing Office, 1950), 1ff.

Chapter 2

[1]C. B. McCullough to J. E. Mackie, National Lumber Manufacturers [sic] Association, 7 September 1937, Folder Org-7, Bridges—Covered, Office of General Files, Oregon Department of Transportation, Salem.

[2]"McCullough, Boyd," *History of the Reformed Presbyterian Church of America* (n.p., n.d.), 586-87, copy held by the Presbyterian Historical Society, Philadelphia, PA.

[3]"McCullough, Conde B.—Personal History for Alumni Records (7/18/32)," Alumni Affairs Collection, Series 21/7/1, Box 81, Iowa State University Archives, Ames, Iowa.

[4]Ibid.; "Funeral of J. B. M'Cullough" [sic], *The Messenger* (Fort Dodge, Iowa), 26 October 1904; Iowa, "State Census, Webster County, 1895," residence number 281, p. 129; see William E. Tucker and Lester G. McAllister, *Journey in Faith: A History of the Christian Church (Disciples of Christ)* (Saint Louis, MO: The Bethany Press, 1975), 315-25; "Former Resident Dies: Mrs. Lenna McCullough Dies at Ames," *Messenger* (Fort Dodge, Iowa), 24 June 1912; Karen J. Blair, *The History of American Women's Voluntary Organizations, 1810-1960* (Boston: G. K. Hall and Co., 1989), nos. 5 and 488.

[5]Interview with John R[oddan]. McCullough, by Louis F. Pierce, Salem, Oregon, 13 May 1980, transcript, 1, 27. It appears from a collection of city directories for Fort Dodge, Iowa, housed at the Fort Dodge Public Library, that the John B. McCullough family moved to that city shortly before 1894. A "J. B. McCullough" was listed in each volume for 1894, 1896, 1898, 1902, 1903, and 1904. In addition, while John B. McCullough was reputably a physician, his name was nowhere found in special listings for doctors of medicine. There was no note whatsoever of an occupation for him in the directories for 1894, 1896, 1898, 1903, or 1904. In the 1902 volume by R. L. Polk and Company, he was listed as a "trav. agt.," or travelling agent, according to an index to abbreviations used in the entries. In 1904, Conde B. McCullough was listed as a "solr.," or solicitor, probably a synonym for an agent or salesman. On J. B. McCullough's death, see Certificate of Death for John Black McCullough, 25 October 1904, No. 94-0038, State Board of Health, Iowa; and "Funeral of J. B. M'Cullough," *Messenger* (Fort Dodge, Iowa), 26 October 1904. Lenna McCullough died in Ames in June 1912. See "Former Resident Dies: Mrs. Lenna McCullough Dies at Ames," *Messenger* (Fort Dodge, Iowa), 24 June 1912.

[6]C. S. Nichols, *Iowa State College of Agriculture and Mechanic Arts Directory of Graduates of the Division of Engineering* (Ames, 1912), hereafter as *ISC Directory*, 39-40.

[7]Almon Fuller, *A History of Civil Engineering at Iowa State College* (Ames: Alumni Achievement Fund of Iowa State College, 1959), 16, 22-23; "Civil Engineering Course," in *Bulletin* (1906), Iowa State College of Agriculture and Mechanic Arts [ISC], 153-67; Anson Marston, "Demands of the Times as to Engineering Education," in *Proceedings of the Iowa State Teacher's [sic] Association, for 1894* vol. 40, pp. 106-10, reproduced in Herbert J. Gilkey, *Anson Marston: Iowa State University's First Dean of Engineering* (Ames: Iowa State University, College of Engineering, 1968), 94.

[8]Marston, "Demands of the Times as to Engineering Education," 106-10, as reproduced in Gilkey, *Anson Marston*, 94. See also "Civil Engineering Course," *Bulletin* (1906), ISC, 153-167.

[9]"Who's Who Among Ames Men," *Iowa Engineer* 26 (December 1925): 14.

[10]Earle D. Ross, *Democracy's College: The Land-Grant Movement in the Formative Stage* (Ames: The Iowa State College Press, 1942), 77-78.

[11]Waterman Thomas Hewett, *Cornell University: A History*, vol. 2 (New York: The University Publishing Society, 1905), 326-28; Ross, *Democracy's College*, 12 (quote);

Samuel Rezneck, *Education for a Technological Society: A Sesquicentennial History of the Rensselaer Polytechnic Institute* (Troy, NY: Rensselaer Polytechnic Institute, 1968), 46; Morris Bishop, *A History of Cornell* (Ithaca: Cornell University Press, 1962), 113, 52 (quote). See also, John Rae, "Application of Science to Industry," in *The Organization of Knowledge in Modern America, 1860-1920*, ed. Alexandra Oleson and John Voss (Baltimore: Johns Hopkins University Press, 1979), 249-68.

[12] "Fuertes, Estévan Antonio," *National Cyclopædia of American Biography*, vol. 4 (New York: James T. White and Company, 1897), 483; James Gregory McGivern, "First Hundred Years of Engineering Education in the United States (1807-1907)," Ph.D. diss. (Washington State University, 1960), 241; Bishop, *A History of Cornell*, 168; Gilkey, *Anson Marston*, 15.

[13] G. Frank Allen, "Address of the President," *Proceedings of the Society for the Promotion of Engineering Education* 7 (1904): 12; Society for the Promotion of Engineering Education, *Report of the Investigation of Engineering Education, 1923-29* 1 (1930): 545-47, quoted in Ross, *Democracy's College*, 155; McGivern, "First Hundred Years of Engineering Education in the United States," 237; Society for the Promotion of Engineering Education, *Report of the Investigation of Engineering Education, 1923-29*, 1 (1930): 545-47, quoted in Ross, *Democracy's College*, 155; Robert Fletcher, "A Quarter Century of Progress in Engineering Education," in *Proceedings of the Fourth Annual Meeting, held in Buffalo, N.Y., August 20, 21, 22, 1896*, by the Society for the Promotion of Engineering Education (Lancaster, PA: New Era Printing Company, for the Society for the Promotion of Engineering Education, 1897), 38.

[14] "Who's Who Among Ames Men," *Iowa Engineer* 26 (December 1925): 14. See also "Marston, Anson," *National Cyclopædia of American Biography* (New York: James T. White and Company, 1962), 44:96-97.

[15] Earle D. Ross, *A History of the Iowa State College of Agriculture and Mechanic Arts* (Ames: The Iowa State College Press, 1942), 128-29.

[16] Ibid., 197.

[17] Ibid., 206 (quote); Gilkey, *Anson Marston*, 14.

[18] Ross, *Democracy's College*, 155-57; Bishop, *A History of Cornell*, 244, 245, 280.

[19] Fuller, *A History of Civil Engineering at Iowa State College*, 16.

[20] Gilkey, *Anson Marston*, 16; Marston, "Demands of the Times as to Engineering Education," 106-10, reproduced in Gilkey, 94 (quote).

[21] Marston and his fellow members of the Society for Promotion of Engineering Education called for a standard four-year college curriculum in civil engineering. Forty percent of its courses focussed on the specifics of the field while the remaining 60 percent gave students a well rounded liberal education. Marston, "Demands of the Times as to Engineering Education," 106-110, reproduced in Gilkey, 94 (quote). See also McGivern, "First Hundred Years of Engineering Education in the United States," 219-23, 240-41.

[22] Anson Marston, "The Engineering Division," *Iowa State Alumnus* 19 (June 1924): 276.

[23] "Civil Engineering Course," *Bulletin* (1906), ISC, 155, 156, 159, 160; Anson Marston, "A Word To Freshman Engineers," *Iowa Engineer* 21 (October 1920): 4 (quote); Marston, "Demands of the Times as to Engineering Education," 106-10, reproduced in Gilkey, *Anson Marston*, 94 (quote).

[24] Fuller, *A History of Civil Engineering at Iowa State College*, 24-25.

[25]"Civil Engineering Course," *Bulletin* (1906), ISC, 157 (quote).

[26]Nichols, *ISC Directory*, 39-40.

[27]"Civil Engineering Course," *Bulletin* (1906), ISC, 157-67.

[28]Kirkham worked for Waddell on the Chicago Elevated Railway System and on several railroad and highway bridges in the Midwest. He later was Waddell's chief draftsman. He designed bridges for many locations around the world. Kirkham then worked for a number companies. He eventually oversaw construction of many bridges on the Pennsylvania and New York Central railroads. He briefly taught civil engineering at Pennsylvania State College. Before moving to Ames, he was a designing engineer for Andrew Carnegie's American Bridge Company of Ambridge, where he became a specialist on movable bridges. See Nichols, *ISC Directory*, 25.

[29]Kirkham regarded his volume as "a self-explanatory manual [of engineering] . . . for practical men." John Edward Kirkham, *Structural Engineering* 2d ed. (New York: McGraw-Hill Book Company, Inc, 1933), vi (quote from "Preface to the First Edition"). Though he left Iowa State in 1919, Kirkham continued to write on structural engineering, publishing subsequent volumes entitled: *Highway Bridges: Design and Cost* (New York: McGraw-Hill, 1932), and *Reinforced Concrete: Theory and Design* (Ann Arbor, MI: Edwards Brothers, 1941).

[30]Carl W. Condit, *American Building: Materials and Techniques from the Beginning of the Colonial Settlements to the Present*, 2nd edition (Chicago: University of Chicago Press, 1982), 168-76.

[31]Ibid., 174-75, 251; Plowden, *Bridges: The Spans of North America*, 298-99.

[32]Anson Marston, "The Choice of Subjects for Theses," *Iowa Engineer* 11 (April 1911): 315-16.

[33]Ibid.

[34]H. B. Walker and C. B. McCullough, "Effect of External Temperature Variation on Concrete Bridges," (B.S. thesis, Iowa State College of Agriculture and Mechanic Arts, 1910), 18-20.

[35]W. Dean Nelson, Associate Registrar, Iowa State University, Ames, to Robert Hadlow, Pullman, WA, 16 July 1992.

[36]Marsh was born in North Lake, Wisconsin, in 1856. He moved to Iowa eighteen years later to attend the Iowa Agricultural College, in Ames, where he received a Bachelor's degree in Mechanical Engineering in 1882. He worked for the Iron Bridge and Manufacturing Company of Cleveland, Ohio, at its Des Moines office. He designed and marketed steel truss spans for the firm whose products rivaled those of the preeminent metal bridge manufacturers. These included the Phoenix Bridge Company's patented iron spans and Keystone Bridge Company's steel bridges. Nichols, *ISC Directory*, 112; Plowden, *Bridges: The Spans of North America*, 65-66.

[37]U.S. Patent Office, Letters of Patent, Reinforced Arch-Bridge, James B. Marsh, 6 August 1912, no. 1,035,026; "Marsh Arch Bridges as a Part of Kansas' Transportation History," *Kansas Preservation* 5 (March-April 1983): 1-5. Marsh hoped to end Daniel Luten's Pennsylvania-based company's dominance over the reinforced-concrete bridge construction market in the Middle Atlantic states and the Midwest. See Emory L. Kemp, *West Virginia's Historic Bridges* ([Charleston]: West Virginia Department of Culture and History, West Virginia Department of

Highways, 1984), 132; and P. A. C. Spero and Co., *Delaware Historic Bridges Survey and Evaluation* ([Dover]: Delaware Department of Transportation, Division of Highways, 1991), 91-93.

[38]Ibid.; "Rainbow Arch Bridge Adds Variety to Kansas Highways," *Kansas Preservation* 2 (November–December 1980): 1-2. Marsh's reinforced-concrete span was not as vulnerable to collapsing from high water as were others because he included features that made them less easily undermined by rushing water. A system of expansion joints throughout the structures prevented the pressures of frost from tearing apart the components of the spans. Finally, unlike many steel truss bridges of the period, Marsh's rainbow arch did not have a wooden road deck. Instead, his design was completely constructed of reinforcing steel and concrete, so, fire danger to the structure was insignificant.

Chapter 3

[1]For a discussion of the evolution of the good roads movement in the U.S., see Bruce E. Seely, *Building the American Highway System: Engineers as Policymakers* (Philadelphia: Temple University Press, 1987), 11-23.

[2]Ibid., 14-16; *America's Highways, 1776-1976: A History of the Federal-Aid Program*, Federal Highway Administration [FHWA], Department of Transportation (Washington, DC: Government Printing Office [1977]), 80-81, 201-202. See also Wayne E. Fuller, *RFD: The Changing Face of Rural America* (Bloomington: Indiana University Press, 1964), 17-48.

[3]*America's Highways*, FHWA, 64-67, 72-73; Seely, *Building the American Highway System*, 17-18. See also *Annual Report*, Office of Public Road Inquiry, Department of Agriculture (Washington, DC: Government Printing Office, 1901-05). A. Marston, "The State's Responsibility in Road Improvement," *Iowa Engineer* 7 (November 1908): 212-13; Thomas R. Agg and John E. Brindley, *Highway Administration and Finance* (New York: McGraw-Hill Book Company, 1927), 31-33.

[4]"An Act to Create a Highway Commission for the State of Iowa," Chapter 105, Highway Commission, H.F. 371, *Supplemental Code of Iowa*, 30th General Assembly, Iowa, as reproduced in *Bulletin*, Iowa Engineering Experiment Station [IEES], 2 (May 1905): 3-4. For a discussion of the movement in Iowa, see Agg and Brindley, *Highway Administration and Finance*, 31-33. For a good discussion of the national movement, see Seely, *Building the American Highway System*, 11-23. See also John E. Brindley, *History of Road Legislation in Iowa*, Iowa Economic History Series, ed. Benjamin F. Shambaugh (Iowa City: The State Historical Society of Iowa, 1912), 217-23; and Marston, "The State's Responsibility in Road Improvement," 212.

[5]*Annual Report for 1913-14*, Iowa State Highway Commission [ISHC], 7-9; *Bulletin*, IEES, 2 (May 1905): 3-4; Brindley, *History of Road Legislation in Iowa*, 217-18.

[6]MacDonald graduated from ISC, in 1904, with a B.S. degree in Civil Engineering. J. W. Eichinger summarized sixteen years of the ISHC history from office reports and documents that no longer exist in "Iowa's Largest State Job," *Iowa Engineer* 21 (October 1920): 1.

[7]"Iowan Receives Medal of Merit" TMs, press release, Information Service, Iowa State College, 21 September 1946; and "Thomas H. MacDonald" TMs, press release, [Information Service], Iowa State College, 20 March 1953, both in Folder

"MacDonald, Thomas H.," College of Engineering, Box 2, Series 11/5/1, Iowa State University Archives, Ames, Iowa; Fuller, *A History of Civil Engineering at Iowa State College*, 44. The commissioner encouraged MacDonald by proposing that on a half-time basis he instruct on highway construction methods in the ISC civil engineering program. *America's Highways*, FHWA, 176.

[8] As a college professor at the state college, he earned an equal amount. Eichinger, "Iowa's Largest State Job," 1-3, 1 (quote). See Agg and Brindley, 35-41, for a good historical discussion of state-aid acts. While states other than Massachusetts provided funds for road construction, only that state used a highway commission to disperse it. *Annual Report for 1913-14*, ISHC, 7.

[9] *America's Highways*, FHWA, 60-62; Seely, *Building the American Highway System*, 26; *Annual Report for 1913-14*, ISHC, 7. See also Burnham, "Essay," 14-18; and Layton, "The Revolt of the Engineers," 109-33 for general discussions on engineers as professionals and apolitical experts.

[10] MacDonald, Thos. H., "Bridge Patent Litigation in Iowa," *Iowa Engineer* 18 (January 1918): 118; Eichinger, 1-2; Earle D. Ross, *A History of Iowa State College of Agriculture and Mechanic Arts*, 257; Anson Marston, "History and Organization of Engineering Experiment Station," *Bulletin*, IEES, 8 (November 1908): 252-56. The station in Iowa and one in Illinois were the first in the nation established under the Hatch Act of 1890, which promoted both agricultural experiment stations and engineering experiment stations connected with each state's land-grant college.

[11] MacDonald, "Bridge Patent Litigation in Iowa," 119-20.

[12] Eichinger, "Iowa's Largest State Job," 2 (quote).

[13] C. B. McCullough, "Designing Department," *Annual Report for 1913-14*, ISHC, 21; Nichols, *ISC Directory*, 39, 44-45. For examples of standardized spans see ISHC, "Iowa Highway Commission Standard Plans—Concrete Culverts, 1911, Drawn by C. B. Mc[Cullough], Approved by T. H. M[acDonald]," copy in author's possession.

[14] A. Marston, "Legislative Appropriation for the Engineering Division," *Iowa Engineer* 9 (July & September 1909): 151 (quote).

[15] Lauren K. Soth, "He Pulled Iowa Out of the Mud," *Alumnus*, Iowa State College, October 1931, 1-3; White and Ames, "Is a County Engineer Necessary?," 283-90.

[16] White and Ames, "Is a County Engineer Necessary?," 283-90; MacDonald, "Bridge Patent Litigation in Iowa," 120 (quote); Marston, "The State's Responsibility in Road Improvement," 211.

[17] White and Ames, "Is a County Engineer Necessary?," 283-90.

[18] Ibid.; *Annual Report for 1913-14*, ISHC, 8.

[19] "State Control of Highway Bridge Construction," editorial, *Engineering News* 67 (13 June 1912): 1137-38, 1137 (quote).

[20] Eichinger, "Iowa's Largest State Job," 2-3; *Annual Report for 1913-14*, ISHC, 8-9.

[21] Anson Marston, "Highway Engineering in Iowa," *Iowa Engineer* 14 (October 1913): 15-16. See Staff Article, "The Organization and Standards of the Iowa Highway Commission," *Engineering and Contracting* 42 (15 July 1914): 55-58. Of forty-four states that had highway departments in 1914, thirty-one also provided financial aid for road construction. *Annual Report for 1916*, ISHC, 5-6, 10 (quote), 23.

[22] Nichols, *ISC Directory*, 39-40; Marston, "Highway Engineering in Iowa," 15-16.

[23] McCullough, "Designing Department," *Annual Report for 1913-14*, ISHC, 21; C. B. McCullough, "Concrete Highway Bridge Construction as Standardized by Iowa Commission," *Engineering Record*, 7 November 1914, 514-17.

[24]McCullough, "Concrete Highway Bridge Construction as Standardized by Iowa Commission," 514-17.

[25]Ibid.

[26]Ibid.

[27][C. B. McCullough] "The Design of Concrete Highway Bridges with Special Reference to Standardization," *Engineering and Contracting* 43 (24 March 1915): 268-270, 268 (quote).

[28]"Obituary—Earl Foster Kelley," TMs, Research Reports Division, Bureau of Public Roads, Department of Commerce, July 1961; "Civil Engineering Prof. Record," in Alumni Affairs Collection, Series 21/7/1, Iowa State University Archives, Ames, Iowa.

[29]McCullough, "Designing Department," in *Annual Report for 1913-14*, ISHC, 21-22. See E. F. Kelley, "Steel Bridge Standards of Iowa Highway Commission," *Engineering Record* 70 (12 December 1914): 631-32; and Staff Article, "Standard I-Beam and Pile Highway Bridges of the Iowa State Highway Commission," *Engineering and Contracting* 42 (29 July 1914): 102-07.

[30]Walker and McCullough, "Effect of External Temperature Variation on Concrete Bridges," 18-20.

[31]Ibid.

[32]C. S. Nichols and C. B. McCullough, "The Determination of Internal Temperature Range in Concrete Bridges," *Bulletin*, IEES, No. 30, 2 (January 1913) 4-6 passim. See [McCullough], "The Design of Concrete Highway Bridges with Special Reference to Standardization," 268-70; "Internal Temperature Range in Concrete Arch Bridges," *Engineering and Contracting* 40 (12 November 1913): 533.

[33]Ibid.

[34]C. B. McCullough, "Are the Highway Commission Bridges Too Heavy?," *Bulletin*, ISHC, No. 12, December 1914, 3-4; see [McCullough] "The Design of Concrete Highway Bridges with Special Reference to Standardization," 270 (quotes).

[35]McCullough, "Designing Department," *Annual Report for 1913-14*, ISHC, 21; [McCullough] "The Design of Concrete Highway Bridges with Special Reference to Standardization," 270 (quotes). For a discussion of McCullough's preferred bridge types, see C. B. McCullough, "Concrete Highway Bridge Construction as Standardized by Iowa Commission," 514-17. "Internal Temperature Range in Concrete Arch Bridges," *Engineering and Contracting*, 533.

[36][Larry Joachims] "Masonry Arch Bridges of Kansas," TMs, Kansas State Historical Society, Topeka, Kansas, [1980], 4; Thomas MacDonald, "Bridge Patent Litigation in Iowa," *Iowa Engineer* 18 (4 January 1918): 119-24.

[37][Joachims] "Masonry Arch Bridges of Kansas," 4; "Bridge Patents and Litigation," *Annual Report for 1916*, ISHC, 39-51; MacDonald, "Bridge Patent Litigation in Iowa," 122. See also a thorough discussion of the Iowa suit in "Lower Court Rules Against Luten Patent," *Engineering News-Record* 80 (17 January 1918): 144.

[38][Joachims] "Masonry Arch Bridges of Kansas," 4. McCullough provided in this short paper a thorough and logical explanation of the advantages of setting standards for highway bridges. One may read it as a direct rebuttal aimed at Daniel B. Luten and his less than adequate patented bridge designs. See C. B. McCullough, "Are the Highway Commission Bridges Too Heavy?: County vs. State Control," *Service Bulletin*, ISHC, December 1914, 3-5.

[39]MacDonald, "Bridge Patent Litigation in Iowa," 120-21; [C. B. McCullough] "Bridge Patents and Litigation," *Annual Report for 1916*, ISHC, 39-51.

[40]"The Luten Patents on Concrete Construction," *Engineering News* 71 (5 February 1914): 329. McCullough, along with parties interested in all eleven cases even met together in Des Moines, Iowa, on 23 January 1914, to discuss their suits against Luten.

[41]MacDonald, "Bridge Patent Litigation in Iowa," 122-27; [Joachims] "Masonry Arch Bridges of Kansas," 4-5; MacDonald, "Bridge Patent Litigation in Iowa," 119-124; [C. B. McCullough], "Bridge Patents and Litigation," *Annual Report for 1916*, ISHC, 45-46 (quotes), 50-51; MacDonald, "Bridge Patent Litigation," 119-24. See also "Lower Court Rules Against Luten Patent," 144; and for extensive extractions from Judge Wade's opinion, see "Decision of U.S. District Court of Iowa on Certain Luten Patents for Concrete Bridges," *Engineering and Contracting* 49 (23 January 1918): 94-96. Federal appeals courts affirmed previous lower-court decisions in Luten patent cases. For example, see "Luten Patent Decision Upheld in Higher Court," *Engineering News-Record*, 81 (26 December 1918): 1200; and "Appeals Court Sustains Decision Against Luten Patents," ibid. 84 (26 February 1920): 417-18. Nevertheless, Luten remained unwavering in his views. See Russell T. MacFall, attorney for D. B. Luten, to *Engineering News-Record*, as in "Mr. Luten's Attorney Takes Exception," *Engineering News-Record* 84 (17 June 1920): 1220-21.

[42]*Annual Report for 1916*, ISHC, 65, 154-55.

[43]*General Catalogue, 1916-1917*, ISC, 14 (1 April 1916): 68; Iowa State College, Commencement Exercises, "Program," 8 June 1916, 14. See also McGivern, "First Hundred Years of Engineering Education in the United States (1807-1907)," 221-22.

[44]"Conde McCullough Marries," *Messenger* (Fort Dodge, Iowa), 6 June 1913; Return of Marriages in the County of Dallas, to the Secretary of Iowa State Board of Health, for Conde B. McCullough and L. Marie Roddan, 4 June 1913; *Barometer*, Oregon Agricultural College (Corvallis, Oregon), 19 September 1916, 3.

[45]*America's Highways*, FHWA, 176.

[46]Soth, "He Pulled Iowa Out of the Mud," 1-3. "Obituary—Earl. F. Kelley," Department of Commerce. Anson Marston continued as Dean of the ISC Division of Engineering until his retirement in 1937. He remained a member of the ISHC until 1927. He was president of land-grant college associations and the American Society of Civil Engineers. He died in an automobile accident near Tama, Iowa, on 21 October 1949. "Dean Emeritus Marston Is Automobile Wreck Victim," *Daily Tribune* (Ames, Iowa), 22 October 1949; "Marston, Anson," *National Cyclopædia of American Biography*, vol. 4. (New York: James T. White and Co., 1962), 967. John E. Kirkham left Ames in 1919 to become the first bridge engineer for the South Dakota State Highway Commission where he designed all state and county bridges. He ended private bridge companies' grip on county governments when he created standardized bridge plans suited to South Dakota's varied topography. Fredrik L. Quivik and Lon Johnson, *Historic Bridges of South Dakota*, ([Pierre]: South Dakota Department of Transportation, 1990), 15.

[47]*Barometer*, Oregon Agricultural College (Corvallis, Oregon), 19 September 1916, 3 (quote).

Chapter 4

[1] Ernest W. Peterson, "McCullough Honor Guest at Banquet," *Oregon Journal* (Portland, Oregon), 8 October 1935, CF-OSA; Interview, Louis F. Pierce with P. M. Stephenson, 4 June 1980, TMs, 2-3, transcript held by Pierce, Junction City, Oregon; *Barometer*, Oregon Agricultural College (Corvallis), 19 September 1916. Edgecomb wrote a thesis on the design and construction of the Siletz River Highway suspension bridge at Upper Ford, Oregon. See "Candidates for Professional Degrees," in *Program, Commencement Exercises*, Iowa State College, 8 June 1916, 14. See also *Directory of Engineering Alumni of Iowa State College* (Ames: The Iowa Engineer, 1938), 21, where it is noted that Rex E. Edgecomb, by the 1930s, became Chief Engineer for the Public Works Administration at its Omaha, Nebraska, office.

[2] E. T. Reed, "O.S.C. President for 25 Years," *Oregon State Monthly* 11 (June 1931): 3-8.

[3] *Service Bulletin*, ISHC, August 1916, 16 (quote); "State Highway Department," *Sixth Biennial Report of the State Engineer of the State of Oregon* (Salem: State Printing Department, 1916), 52.

[4] *President's Biennial Report for 1912-14*, Oregon Agricultural College [OAC], xvii-xxi.

[5] Ibid. Gordon Skelton called the highway engineering course "essentially a civil engineering course, but with stronger emphasis placed upon purely technical highway engineering subjects." See "Department of Civil and Highway Engineering," *The Student Engineer [OAC]* 9 (1916): 290-92, 291 (quote).

[6] Grant Covell, "Report of the School of Engineering and Mechanic Arts," *President's Biennial Report for 1914-16*, OAC, 97-99; *Annual Catalogue for 1916-17*, OAC, 211-23. For a list of courses and instructors, see *Students Handbook, 1916-17*, OAC, 30.

[7] Covell, "Report of the School of Engineering and Mechanic Arts," 97-99.

[8] *Annual Catalogue for 1917-18*, OAC, 208-09; *Students Handbook, 1917-18, Second Semester*, OAC, 11. For a short history of engineering at the OAC to 1926, see S. M. Dolan, "Early History of Engineering School is Outlined," *Oregon State Technical Record* 3 (May 1926): 3-4, 16, 18-22.

[9] Stephenson interview, 6, 9, 15; interview, Louis F. Pierce with John R. McCullough, 13 May 1980, TMs, 10, transcript held by Pierce, Junction City, Oregon. John R. McCullough was C. B. McCullough's son. He was born in Ames, Iowa, in 1914 and died in Salem, Oregon, in 1984. He worked as an attorney in Oregon state government. Samuel M. P. Dolan, born in Folkstone, Kent, England, on 14 August 1884, earned a Civil Engineering degree at Notre Dame. He began his teaching career at OAC in August 1910. Smith earned a B.S. degree from OAC. He began his teaching career at his alma mater in July 1914. By the late 1920s Smith taught senior structures courses. See *Lectures, Recitations and Laboratory Periods, First Term, 1927-28*, Oregon State Agricultural College, 19. By the early thirties he moved to Salem to work for McCullough. He designed the second Tacoma Narrows suspension bridge, in the 1940s, for the Washington State Toll Bridge Authority.

[10] Ralph Watson, compiler, *Casual and Factual Glimpses at the Beginning and Development of Oregon's Roads and Highways* (Salem: Oregon State Highway Commission, [1951]), 19; Hugh Myron Hoyt, "The Good Roads Movement in Oregon: 1900-1920," Ph.D. diss. (University of Oregon, 1966), 246.

[11]Watson, *Casual and Factual Glimpses*, 29–30; [Charles Purcell] "Preface," 5 (quote), "Investigation of Recent County Bridge Construction," 177–90, and [Purcell] "Bridges Designed and Built by the State Highway Commission," 168–76, *Annual Report for 1913-14*, OSHC.

[12]Bowlby resigned as State Highway Engineer on 31 March 1915 amid allegations of poor management. Hoyt, "The Good Roads Movement in Oregon," 249. With the reorganization, the State Engineer oversaw the work of the highway department until 1916, when the legislature created the modern state highway commission. State Engineer John H. Lewis, in his 1914-16 report, made a strong case for a more powerful state highway commission by describing, in great detail, the plight of taxpayers unknowingly purchasing expensive, poorly constructed bridges. "State Highway Department," *Sixth Report of the State Engineer (1914-16)*, Oregon, 7–24. The state legislature combined his job with that of the State Engineer, who oversaw water resources in the state. See "An Act Abolishing the Office of State Highway Engineer," Chapter 337, *General Laws of Oregon, 1915*, 537; Seely, *Building the American Highway System*, 46–48.

[13]Seely, *Building the American Highway System*, 46–48.

[14]"An Act to Provide a General System of Construction, Improvement and Repair of State Highways. . . ," Chapter 237, *General Laws of Oregon, 1917*, 447–57; "An Act to Provide for the Construction of Roads and Highways in the State of Oregon. . . ," Chapter 423, *General Laws of Oregon, 1917*, 897–905; Watson, *Casual and Factual Glimpses*, 28; OSHC, *Biennial Report for 1917-18*, 7–8; Hoyt, "The Good Roads Movement in Oregon," 230–31.

[15]For projects on post roads, involving Federal-Aid Road Act dollars, the OSHD prepared plans and let contracts for work on bridges and roads subject to the OPR's approval. *Biennial Report for 1917-18*, OSHC, 19.

[16]Later, Purcell became a field engineer for the OPR's Portland office before becoming an engineer for the California Department of Highways. In the 1930s he oversaw the design and construction of the San Francisco-Oakland Bay Bridge. "Report of the State Highway Engineer," *Biennial Report for 1917-18*, OSHC, 17–25. Howard Holmes later oversaw bridge construction for the Montana Department of Highways and became its State Highway Engineer in 1941. "Background of a State Highway Engineer," *Pacific Builder and Engineer*, July 1941, 65.

[17]See "Table B—Bridges and Culverts," *Biennial Report for 1917-18*, OSHC, 54–57; *Biennial Report for 1917-18*, OSHC, 38.

[18]Watson, *Casual and Factual Glimpses*, 29; "An Act to Provide for the Construction of Roads and Highways in the State of Oregon. . . ," Chapter 173, *General Laws of Oregon, 1919* 241–49; "An Act to Provide for the Construction of a Highway to be Known as 'The Roosevelt Coast Military Highway'. . . ," Chapter 345, *General Laws of Oregon, 1919*, 610–13; "An Act to Provide a License Tax on Gasoline, Distillate, Liberty Fuel and Other Volatile and Inflammable Liquids . . . for the Purpose of Operating or Propelling Motor Vehicles," Chapter 159, *General Laws of Oregon, 1919*. See Edmund P. Learned, "Gasoline Taxes: Theory, Practice, and Hazards," *Engineering News Record* 104 (2 January 1930: 12–16; and John Chynoweth Burnham, "The Gasoline Tax and the Automobile Revolution," *Mississippi Valley Historical Review* 48 (December 1961): 435–59, especially 437–40.

Hoyt also discussed political aspects of the fuel tax issue, see "The Good Roads Movement in Oregon," 231–39.

[19]Watson, *Causal and Factual Glimpses*, 29–30.

[20]Anson Marston, "What is Engineering," *Iowa Engineer* 21 (May 1921): n.p.; A. Marston, "The Civil Engineer and His Place in the World's Economy," *Iowa Engineer* 9 (November 1909): 196–202, 199, and 201 (quotes).

[21]See Chapter 3; *Barometer*, Oregon Agricultural College (Corvallis), 15 April 1919.

[22]The swiftness of McCullough's efforts in assembling his staff at the OSHC bridge department attested to his commitment to designing and building bridges. Minutes from the OAC Board of Regents quarterly meeting are located at the Oregon State University Archives. There were no entries mentioning McCullough during the period of his employment, 1916 through 1919. The Board often convened outside of its scheduled quarterly meeting. Dwight A. Smith, *Columbia River Highway Historic District: Nomination of the Old Columbia River Highway in the Columbia Gorge to the National Register of Historic Places* (Salem: Environmental Section, Technical Services Branch, Oregon State Highway Division, Oregon Department of Transportation, 1984), 60; Plowden, *Bridges: The Spans of North America*, 318–20.

[23]"Report of the State Highway Engineer," *Biennial Report for 1919-20*, OSHC, 28.

[24]The fifth classmate declined McCullough's offer and sought employment in California. See Stephenson interview, 1–2.

[25]Stephenson interview, 1–2. All four men were active in the Civil Engineers' Association at the OAC. Archibald, Stephenson, and Ricketts, like McCullough, were also members of Sigma Tau, an engineering honorary. See "Raymond Archibald", "Ellsworth Ricketts," "Mervyn Stephenson," in *The Beaver [by the Junior Class (Class of 1920)]* (Corvallis: Oregon Agricultural College, 1919), 40, 50, 51. Skelton was away from OAC for over a year as an officer in the U.S. Marine Corps during World War I. See "Albert G. Skelton," *The Beaver [by the Junior Class (Class of 1918)]* (Corvallis: Oregon Agricultural College, 1917), n.p. See "'Peany' is Scholar and Varsity 'O' Man," *Barometer*, Oregon Agricultural College (Corvallis), 28 May 1919. Mervyn Stephenson recalled that McCullough "demanded perfection in our work. . . . I know that I wasn't too good a draftsman." He "came along and looked at my work and said 'You are going to have to do better or I'm going to send you out in the field' and not two weeks later I was on my way [to La Grande]." Stephenson interview, 3 (quote here).

[26]*Biennial Report for 1921-22*, OSHC, 77; C. S. Nichols, *ISC Directory*, 72. See listing of Rosencrans in *Schedule of Lectures, Recitations and Laboratory Periods, 1918-19, Third Term*, OAC, 7. For background on Reeves, see "National Intelligence Committee Application for William Alfred Reeves," ISC, 3 April 1917; and "W. A. Reeves, State Highway Engineer Dies," *Capital Journal* (Salem), [1951], clipping, in Alumni Files, Series 21/7/1, Iowa State University Archives, Ames.

[27]See Chapter 3. The OPR became the BPR in 1918.

[28]Mississippi Valley Conference of State Highway Departments, "Historical Highlights, 1909-1974," TMs, copy located in Iowa Department of Transportation Library, Ames, pp. x-38, x-39, x-40, and x-41. The AASHO was formed in 1914 to "provide mutual cooperation and assistance to the State Highway departments of the several States and the Federal Government. . . ." It worked closely with the

BPR as a private advisory organization in setting nationwide standards for several issues, including road materials, bridge specifications, and highway signage. At present, the organization, now the American Association of State Highway and Transportation Officials (AASHTO), serves much the same purpose. See American Association of State Highway Officials, *AASHO: The First Fifty Years, 1914-1964* (Washington, DC: 1965), 53. At the beginning of fiscal year 1920, on 1 July 1919, MacDonald's job was retitled "Chief" of the BPR. For a good discussion of the events that transpired in choosing MacDonald as Chief of the BPR, see Seely, *Building the American Highway System*, 42-62.

[29]Seely, *Building the American Highway System*, 56 (quote); U.S. Department of Agriculture, Bureau of Public Roads, *Annual Report for 1919*, 397-98 (quote).

[30]"Bridge Department," *Biennial Report for 1919-20*, OSHC, 49; "Bridge Department," *Biennial Report for 1921-22*, OSHC, 60; "Bridge Department," *Biennial Report for 1923-24*, OSHC, 59-61.

[31]C. B. McCullough, *Economics of Highway Bridge Types* (Chicago: Gillette Publishing Co., 1929), 1-2 (quotes).

[32]All plans for bridges over navigable streams, bays, and inlets required the U.S. War Department's, and later the Coast Guard's, approval for meeting minimum vertical and horizontal standards for ship clearance. Ibid., 3-5, 10-16.

[33]McCullough, *Economics of Highway Bridge Types*, 23. McCullough's views were evident in at least one OAC student's writings. See Milton Harris, "Types of Concrete Highway Spans," *The Student Engineer*, OAC, 10 (1917): 23-26.

[34]McCullough, *Economics of Highway Bridge Types*, 22, 23 (quote).

[35]C. B. McCullough, "How Oregon Builds Highway Bridges," *Oregon Motorist*, February 1930, 14-15, 27; McCullough, *Economics of Highway Bridge Types*, 23 (quote).

[36]McCullough, *Economics of Highway Bridge Types*, 154-171.

[37]William G. T'Vault, who settled in the Sams Valley in the 1850s, called the rolling hills of area around Rock Point the Dardanelles, for it reminded him of a region in Europe of that name. In 1859, pioneers established a post office near pointed rocky outcroppings on the Rogue River that flows through the valley. They named it "Rock Point" for the geological formations. See Lewis A. McArthur, *Oregon Geographic Names*, 6th ed., revised by Lewis L. McArthur (Portland: Oregon Historical Society, 1992), 236, 716. A ferry crossed the Rogue near Rock Point. It provided service from the 1850s until a simple bridge replaced. By the 1870s another early Oregonian, Thomas Chavner, built his "Centennial Bridge" a few miles up river from Rock Point. There, he platted the town of Gold Hill. In the early 1880s, the Oregon and California Railroad looked at both towns to find a permanent railway station in the region for its line. See Engineering Antiquities Inventory," Environmental Section, Technical Services Branch, ODOT, November 1982; and "Fact Sheet Eight: Rock Point Bridge" (Jacksonville: Southern Oregon Historical Society, n.d.) [1].

[38]"Rock Point Bridge (No. 332A)," Job Records Cards, BS, TS, ODOT; "Two Interesting Concrete Bridges in Oregon," *Engineering and Contracting*, 26 October 1921, 389. On 10 June 1919, the OHSC approved a bid by Parker and Banfield, a Portland contractor, to build the Rock Point Arch. "Description of Work of the State Highway Department in the Counties of the State: Jackson County, Rock Point Arch," *Biennial Report for 1919-20*, OSHC, 265 (quote).

[39] Arch ribs measured four feet wide and twenty-six inches deep at the crown, and seventy-nine inches deep at the skewbacks. See "Bridge Plans," Rogue River Bridge at Rock Point (No. 332A), Drawings No. 835, 838; and "Bridge Log" for the Highways of Oregon, BS, TS, ODOT. Originally, McCullough planned to fasten the ribs into the deep chasm's walls through key openings in the solid rock. Yet, during excavation, crews found that fissures had made it uncertain whether this type of anchorage was substantial enough to contain the horizontal thrust of the rib ends. McCullough excavated beyond the original keys and grouted in place two 2-1/2" swedge bolts at the ends of each rib. These were five feet in length, and along with reinforcing bars of the ribs, hooked into 2-1/2" transverse "anchor bars" at the skewbacks. He determined the required size of these rods by calculating the bending strain that wet concrete created through horizontal thrust upon the rib bars. See C. B. McCullough, "Arch Bridge Ribs Anchored by Concrete Keys," *Engineering News-Record* 83 (27 November–4 December 1919): 924.

[40] "Two Interesting Concrete Bridges in Oregon," *Engineering and Contracting*, 26 October 1921, 391; R. P. Clark to Parker and Banfield, 28 July 1919, "Rock Point Bridge (No. 332A)," Microfilmed Records, BS, TS, ODOT; McCullough worried that the truss bridge might settle under the load of the wet concrete, so he had all members of the wooden structure drawn up tight. The design allowed for a deflection of 1-3/4" under an entire load of wet concrete. Yet, in practice, the average drop was only 1/2" between the ribs. C. B. McCullough, "Truss Centering Used for 113-Ft. Concrete Arch," *Engineering News-Record* 84 (29 April 1920): 851–52.

[41] "Two Interesting Concrete Bridges in Oregon," *Engineering and Contracting*, 26 October 1921, 391. Of the bridge's total cost, Jackson County paid $23,000. The state assumed the balance. "Description of Work of the State Highway Department in the Counties of the State: Jackson County, Rock Point Arch," *Biennial Report for 1919-20*, OSHC, 265 (quote). Other large concrete bridges that McCullough completed in 1920 include the Fifteen Mile Creek (Seifert) Viaduct, the Mill Creek (West Sixth Street) Bridge, and the Mosier Creek Bridge, all in Wasco County; and the Oswego Creek Bridge in Clackamas County.

[42] Carl Price Richards, "Design and Construction of the Bridges," *Souvenir Program: Oregon City-West Linn Bridge* (n.p.: 1982), held by ES, TS, ODOT; "Description of Work of the State Highway Department—Oregon City Bridge Design," *Biennial Report for 1919-20*, OSHC, 161; "Report of the State Highway Engineer—Bridges," and "Description of Work of the State Highway Department—Oregon City Bridge," *Biennial Report for 1917-18*, OSHC, 21, 75; C. B. McCullough, "Large Steel Arch Ribs Encased in Gunite," *Engineering News-Record* 88 (8 June 1922): 942–45.

[43] McCullough, "Large Steel Arch Ribs Encased in Gunite," 942–43.

[44] C. B. McCullough, "Old Suspension Bridge Used in Erecting New Arch," *Engineering News-Record* 89 (2 November 1922): 730–33; McCullough, "Large Steel Arch Ribs Encased in Gunite," 942–45.

[45] The Historic American Engineering Record, in the summer of 1990, created measured drawings and written documentation of the design and construction of twenty-six Oregon bridges. This was done as part of its mission to record though drawings, words, and photographs, examples of the country's industrial and

engineering. Many of the Oregon bridges surveyed as part of the project were McCullough-designed structures. Kenneth J. Guzowski, "Willamette River Bridge, 1922, Clackamas County, Oregon, HAER No. OR-31," Oregon Historic Bridges Recording Project, Summer 1990, Historic American Engineering Record [HAER-1990], National Park Service, deposited in the Library of Congress, Washington, DC, 1-9. McCullough, "Old Suspension Bridge Used in Erecting New Arch," *Engineering News-Record* 89 (2 November 1922): 730-33; C. B. McCullough, "Gunite Retains Integrity on Oregon Road Bridges," *Engineering News-Record* 111 (31 August 1933): 259-60.

[46]C. B. McCullough to J. C. Ainsworth, 23 November 1922, copy held by Ray Allen, Portland. McCullough referred to the following editorials, "A Novel Bridge Design," *Engineering News-Record* 88 (8 June 1922): 940; and "Original Bridge Thought," *Engineering News-Record* 89 (2 November 1922): 727.

[47][C. B. McCullough] "The Design of Concrete Highway Bridges With Special Reference to Standardization," *Engineering and Contracting* 43 (24 March 1915): 270.

[48]Newspaper accounts in December 1861 reported a flood carrying away a bridge at this location. Presumably, local citizens rebuilt it. By 1887, because of increase road traffic on the road, by then part of the Oregon-California Stage Route, Douglas County called for bids on a new "combination cantilever steel bridge" at the same location. Finally, a steel triple through-arch structure replaced this span in 1912. See *Oregon Statesman* (Salem), 16 December 1861; *Review* (Roseburg, Oregon), 7 October 1887 and 25 January 1889; Stephen Dow Beckham, *Land of the Umpqua: A History of Douglas County, Oregon* (Roseburg, OR: Douglas County Commissioners, 1986), 195. The OSHC awarded the contract on 25 July to Portland builder, H. E. Doering. Work commenced on 15 August and nearly two years later, on 26 April 1924 the bridge was completed. On 27 April, the state dedicated the structure to a recently retired state highway commissioner, Robert A. Booth. See "Robert A. Booth Bridge, Douglas County," in "Description of Work of the State Highway Department in the Counties of the State," *Biennial Report for 1921-22*, OSHC, 263.

[49]McCullough, *Economics of Highway Bridge Types*, 103-06.

[50]C. B. McCullough, "How Oregon Builds Highway Bridges," *Oregon Motorist* 10 (February 1930): 13-15, 27.

[51]*Biennial Report for 1923-24*, OSHC, 265-66 and *Biennial Report for 1925-26*, OSHC, 293; McCullough, *Economics of Highway Bridge Types*, 103-06. The Winchester Bridge over the North Umpqua River consisted of four 10'-2" deck girder spans, seven 112-foot ribbed deck-arches, and two 10'-2" deck girders, all of reinforced concrete. Total length of the structure, including 20'-4" abutments at each end, was 784 feet. Deck to streambed distance averages 56 feet. Roadway width was 19'-4". Milepoint 12.21 on Oakland Shady Highway No. 234, in "Bridge Log of Oregon Highways," BS, TS, ODOT; "Bridge Plans," Winchester Bridge (No. 839), BS, TS, ODOT, Drawings 1771-1789.

Chapter 5

[1]Siegfried Giedion, *Space, Time and Architecture* (Cambridge: Harvard University Press, 1967), 467.

[2][List of Personnel], *Annual Report of Bridge Department for Year Ending Nov. 30, 1923*, OSHC, copy at Bridge Section library, BS, TS, ODOT.

[3]"Report of State Highway Engineer," *Biennial Report for 1925-26*, OSHC, 75; "Report of State Highway Engineer," *Biennial Report for 1927-28*, OSHC, 70; "Report of State Highway Engineer," *Biennial Report for 1929-30*, OSHC, 80; *Biennial Report for 1931-32*, OSHC, 31-32.

[4]Thomas H. MacDonald, "Highway Administration in the United States," *Good Roads* 68 (November 1925): 277 (quote).

[5]McCullough, *Economics of Highway Bridge Types*, 57-82, especially 69-70.

[6]Creosote treating of timbers for bridges was not widely used in Oregon before the early 1930s because the process of impregnating the wood with chemical preservatives was not locally available and meant the need to import treated lumber from the state of Washington. Oregon woodsmen voiced opposition to the use of "foreign" timber for state-sponsored construction projects. C. B. McCullough, "Timber Highway Bridges in Oregon," *Engineering News-Record* 109 (25 August 1932): 213-14.

[7]McCullough, *Economics of Highway Bridge Types*, 80-83, 168-70.

[8]Oregon, along with all other states, began in the 1920s to follow guidelines for road alignments, pavement specifications, and bridge standards that the American Association of State Highway Officials prepared through the BPR's guidance. The AASHO's membership included high-ranking engineers from each state's highway engineering staff who joined together to promote road building activities across the country. The federal agency, as a participant, set the minimum standards for many highway construction activities based on its own ongoing research program. It published them through the AASHO to soften any criticism of it exerting dominance as a national entity over the federal make up of the country's highway building enterprise. Bruce S. Seely, *Building the American Highway System*, 122-25; McCullough, *Economics of Highway Bridge Types*, 34-39, 162-64, 162 (quote).

[9]See chapter 4 for a discussion of the evolution of McCullough's philosophy.

[10]Dwight A. Smith, James B. Norman, and Pieter T. Dykman, *Historic Highway Bridges of Oregon*, 2nd ed. (Portland: Oregon Historical Society Press, 1989), 281; *Biennial Report for 1927-28*, OSHC, 319.

[11]See Robert W. Hadlow, "Gold Hill Bridge over the Rogue River, 1927, Jackson County, HAER No. OR-37," HAER-1990.

[12]C. B. McCullough, "Oregon Steel Arch Bridge Erected By Steel Cableway," *Engineering News-Record* 96 (13 May 1926): 760-62.

[13]Conde B. McCullough and Edward S. Thayer, *Elastic Arch Bridges* (New York: John Wiley and Sons, 1931), 18-19; C. B. McCullough, "Steel Arch Bridges—General," in George A. Hool and W. S. Kinne, editors-in-chief, *Movable and Long-Span Steel Bridges* (New York: McGraw-Hill Book Co., 1923), 359-74.

[14]Ibid.

[15]"Description of Work of the State Highway Department," *Biennial Report for 1925-26*, OSHC, 316, 323-24, 317 (quote); Smith, Norman, and Dykman, *Historic*

Highway Bridges of Oregon, 100; Lawrence C. Merriam and David G. Talbot, *Oregon's Highway Park System 1921-1989: An Administrative History* (Salem: Oregon Parks and Recreation Department, 1992): 211-12, 261-62.

[16]C. B. McCullough to Robert W. Sawyer, 16 May 1927, Box 4, Correspondence of the Chief Counsel, 68A-34-1, HDR, OSA.

[17]C. B. McCullough to J. M. Devers, 16 May 1927, Box 4, Correspondence of the Chief Counsel, 68A-34-1, HDR, OSA.

[18]For discussions of the Highway Research Board, see Highway Research Board [HRB], *Ideas and Actions: A History of the Highway Research Board, 1920-1970* (Washington, DC: National Research Council, National Academy of Sciences, [1971]), 16-18; and Seely, *Building the American Highway System,* 109-12.

[19]See C. B. McCullough and Phil A. Franklin, "Bascule Bridges"; and C. B. McCullough, "Steel Arch Bridges—General," "Analysis of Three-hinged Arch Bridges," Analysis of Fixed Arches," and "Analysis of Two-hinged Arches," in Hool and Kinne, *Movable and Long-Span Steel Bridges,* 1-157, 359-74, 375-92, 393-482, 483-89. See also David B. Steinman's chapters, "Continuous Bridges," and "Suspension Bridges," 199-255, 256-88. Steinman received praise for a previous work, *A Practical Treatise on Suspension Bridges: Their Design, Construction and Erection* (New York: John Wiley and Sons, 1922).

[20]C. B. McCullough, "Maintaining Oregon's Highway Bridges: The Methods Used by the Oregon State Highway Department Described in a Paper Presented at the Recent Meeting of the American Association of State Highway Officials," *Roads and Streets* 67 (March 1927): 115-20. McCullough's two Department of Agriculture publications were, *Highway Bridge Location,* Department Bulletin No. 1486, published in 1927; and *Highway Bridge Surveys,* Technical Bulletin No. 55, published in 1928. The Yugoslavian government in 1930 issued a Serbian language translation of *Highway Bridge Location* as McCullough, C. B. *Izbor Polozaja Drumskih Mostova.* Uz saradnju Biroa za drzvae putove Ministarstva poljoprivrede Sjedinjenih Drzava. Prevod sa engleskog. Beograd, Stampano u grafickoj radionici Min. gradevina, 1930. McCullough, *Economics of Highway Bridge Types,* iii (quotes).

[21]McCullough, *Economics of Highway Bridge Types,* iii and 2 (quotes).

[22]"State Highway Engineer's Report," *Biennial Report for 1919-20,* OSHC, 63-64.

[23]Joseph McClellan Devers, a former Lane County District Attorney, came to Salem to fill the post. "J. M. Devers, Highway Attorney, Summoned," *Capital Journal* (Salem), 1 October 1951; "J M Devers," *Oregon Voter,* 5 May 1934, p. 16. C. B. McCullough to J. M. Devers, 15 December 1920, in "Bridge Laws," Box 4, Correspondence of Chief Counsel J. M. Devers, 68A-34-1, HDR, OSA, four-page letter, 4.

[24]C. B. McCullough to J. M. Devers, 15 December 1920, in "Bridge Laws," Box 4, Correspondence of Chief Counsel J. M. Devers, 68A-34-1, HDR, OSA, two-page letter.

[25]Interview, Robert W. Hadlow with Richard Devers, son of J. M. Devers, June 1992, Salem, Oregon, notes held by author.

[26]For course descriptions see "Annual Catalog," *Bulletin,* Willamette University [WU], 19 (April 1926): 88-90. A 1950s fire destroyed all records of the WU Office of the Registrar, including all records of student course work, transcripts, etc. The only record of McCullough's enrollment, besides his diploma, is his inclusion in the

annual listing of all undergraduates and law students at the end of each year's catalog. Even here, though, McCullough's name did not consistently appear with accuracy. In the *Bulletin, WU*, 19 (April 1926): 102, he was listed as "C. B. McCullough," a freshman student, School of Law. The next year, in the *Bulletin, WU*, 20 (April 1927): 108, he was listed as "Conda McCullough," a junior student, School of Law. Finally, in the *Bulletin, WU*, 21 (February 1928): 113, he was listed as "William McCullough," senior student, School of Law. The *Bulletin, WU*, 22 (February 1929): 106, listed "Conde B. McCullough," as having received a "Bachelor of Laws" degree in the spring 1928 commencement.

[27]Conde B. McCullough and John R. McCullough, *The Engineer at Law: A Resumé of Modern Engineering Jurisprudence*, Forewords by Hon. James T. Brand and Hon. J. M. Devers (Ames: Iowa State College Press, issued by the Collegiate Press, Inc., in cooperation with the Oregon State Highway Department, 1946), 15, 16, and 26-27 (quotes). McCullough, in views expressed in the quotations, carried on a progressive ideal of unselfish service to society. See Burnham, "Essay," 20.

[28]Federal Writers' Program of the Work Projects Administration, *Oregon: End of the Trail*, American Guide Series (Portland: Binsford and Mort, 1940), 363; "Report of State Highway Engineer" and "Description of Work of the State Highway Department," *Biennial Report for 1925-26*, OSHC, 71-72, 369; *Biennial Report for 1927-28*, OSHC, 371-72; *Biennial Report for 1929-30*, OSHC, 262-63; Smith, Norman, and Dykman, *Historic Highway Bridges of Oregon*, 101-02.

[29]*Biennial Report for 1927-28*, OSHC, 66 (quote); *Biennial Report for 1929-30*, OSHC, 13-14.

[30]"Report of State Highway Engineers," *Biennial Report for 1927-28*, OSHC, 72-73.

[31]See various letters, File 8-PF1; Robert L. Withrow, Editor, *Curry County Reporter* (Gold Beach, Oregon), to H. B. Van Duzer, Highway Commissioner, 15 December 1928, File 8-4; and J. G. Eckman to Oregon State Highway Commission, 13 August 1929, and attached undated newspaper article in File 8-4, 76A-90/3, HDR, OSA.

[32]The OSHC chose from among eight designs proposing structures at different points along the river. The bridge selected, costing an estimated $628,000 was the most costly of the group but required no expensive realignment of the coast highway at its approaches. The commission called for bids for the seven-span reinforced-concrete deck-arch structure at its December 1929 meeting. Its action attracted attention from many large construction firms. On 16 January 1930, the Mercer-Fraser Company of Eureka, California, received the contract for $568,181. "Preliminary Estimate [1927]," File 8-4, 76A-90/3, HDR, OSA. See Betty Van Leer, "Spanning the Mighty Rogue—How the Bridge was Built," Rogue Coast Supplement to the *Curry County Reporter* (Gold Beach, Oregon), 26 May 1982, 7-9.

[33]Van Leer, "Spanning the Mighty Rogue," 8.

[34]Albin L. Gemeny and C. B. McCullough, *Application of Freyssinet Method of Concrete Arch Construction to the Rogue River Bridge in Oregon, A Cooperative Research Project by the U.S. Bureau of Public Roads and the Oregon State Highway Commission* (Salem: Oregon State Highway Commission, 1933), 2. In the early twentieth century, French bridge engineer Eugène Freyssinet perfected a method for decentering reinforced-concrete bridge arches that seemed quite unorthodox at the time.

Freyssinet was born in 1879 and studied at the École des Ponts et Chaussées
(College of Bridges and Roads) at the turn of the century. He sought to subject
concrete to an initial compression to neutralize tensile stresses and applied the
method as early as 1907, on a 100-foot arch over the Bresle River. Nevertheless, it
did not gain use in the United States for over two decades. See H. P. Hopkins, *A
Span of Bridges: An Illustrated History* (New York: Praeger, 1970), 261-262, and J. T.
Thompson, "Freyssinet Method of Arch Construction," *Baltimore Engineer* 5
(January 1931): 4-6. See also Robert W. Hadlow, "Oregon's Isaac Lee Patterson
Bridge: The First Use of the Freyssinet Method of Concrete Arch Construction in
the United States, 1932," *IA (The Journal of the Society for Industrial Archeology)* 16,
No. 2 (1990): 3-14.

[35]Hopkins, 262-67. Freyssinet attempted to compensate for shrinkage of concrete,
dead load, differential temperature changes, movement of supports, and elastic and
plastic shortening. See Gemeny and McCullough, *Application of Freyssinet Method of
Concrete Arch Construction*, 4-5; Conde B. McCullough and Albin L. Gemeny,
"Designing the First Freyssinet Arch to Be Built in the United States," *Engineering
News-Record* 107 (26 November 1931): 841; McCullough and Thayer, *Elastic Arch
Bridges*, 337-38. See also E. Freyssinet, "The Bridge at Villeneuve-sur-Lot,
Department of Lot and Garonne, France: Advances in the Construction of Great
Arches," trans. Lloyd G. Frost, TMs, [1925], pp. 28-29, original appeared in *Le
Génie Civil*, 79 [1921].

[36]Ibid.

[37]"Isaac Lee Patterson Bridge," *House Concurrent Resolution No. 1*, 24 February 1931,
Legislature, 36th Assembly, Oregon; and OSHC to Mrs. H. J. Edwards, Coos Bay
Chapter of the Daughters of the American Revolution, January 18, 1932, File 8-4,
76A-90/3, HDR, OSA. The roadway measured 27 feet, with a total structural
width of 34 feet. The approach viaducts rested on bents anchored to solid rock, as
did the abutments, piers 1 and 8, at the extreme north and south ends of the main
structure. Piers 2 through 7 were solid pedestals resting on piling. Footings for
piers 2, 4, 5, and 7 consisted of 180 timber piles driven vertically. Bases measured
29 by 38 feet. Concrete seals, 8 to 10 feet thick, capped the pile heads and served
as foundations for the piers. McCullough designed pedestals 3 and 6 as abutment
piers to resist heavy thrust from adjacent arches. They rested on grids of 260 batter
piles, driven half each direction along the longitudinal centerline of the span, at
angles 20 degrees off the vertical centers of the piers. Crews used hoists and
derricks operating clamshell buckets to excavate piers. They drove the piling with
a Vulcan steam hammer, using a five-ton weight, which delivered 60 blows per
minute. By 1 December 1930, they had poured concrete for all of the piers. See
"Bridge Plans," Bridge at Mouth of the Rogue River (No. 1172), BS, TS, ODOT,
Drawings 3875, 3878, 3879, 3888, and 3889. See W. A. Scott, "Rogue River Bridge
at Gold Beach, Oregon," *Western Construction News and Highways Builder*, 25 May
1932, 281.

[38]Another term for falsework is centering. When it and forms are removed during
construction the process is called striking the centering or decentering. Gemeny
and McCullough, *Application of Freyssinet Method of Concrete Arch Construction*, 7-8.

[39]Ibid. See Gemeny and McCullough in its entirety for a complete discussion of the
Freyssinet method and its application at the bridge.

[40]McCullough wrote to the Paris Établissement Morane Jeune in June 1930 for plans for its 250 metric-ton hydraulic jacks, and necessary valves and pumps (250 metric tons = 275.6 English tons). The company replied in late July with drawings for a forged steel model that had been used on the arch crowns of the La Tournelle bridge in Paris in 1928. A second model, differing from the first in that its piston was threaded, prevented the jack from accidentally compressing if it lost hydraulic pressure. McCullough chose the latter version. See R. H. Baldock to Ian Macallan, 9 September 1952, Bridge at Mouth of Rogue River (No. 1172), Maintenance Files, BS, TS, ODOT. McCullough placed the order for the jacks in late September 1930. The shipment from the Morane Company arrived by steamer in Washington, D.C., in January 1931, but did not reach the Pacific Northwest for another month. The Bureau of Public Roads' Division of Tests held them for the interim to calibrate the pumps and gauges and then sent the equipment, by rail, to Salem. The invoice from Paris totaled $5,950. By adding charges for customs duties, brokers' fees, and freight, the shipment cost about $8,200. [Établissment Morane Jeune] to [McCullough], July 24, 1930; McCullough to Gemeny, 9 August 1930; Morane Hydro to McCullough, telegram, 26 September 1930; Consumption Entry, United States Customs Service, District No. 13, Invoice No. 32,176, 13 January 1931; Receipt, L. P. Seibold, Inc., Washington, D.C., 22 January 1931; Freight Bill No. 5,378, Baltimore & Ohio Railroad Co., 20 January 1931; McCullough to E. S. Thayer, OSHC, 29 August 1931, all from Bridge No. 1172, Microfilmed Records, BS, TS, ODOT. For a description of construction of the arched bridge at Plougastel, see E. Freyssinet, "The 600-ft. Concrete Arch Bridge at Brest, France," trans. by S. C. Hollister, *Proceedings of the American Concrete Institute* 25 (1929): 83–97, and discussion, 98–99.

[41]Gemeny and McCullough, *Application of Freyssinet Method of Concrete Arch Construction*, 8–9; McCullough reasoned that as the deck and spandrel columns were placed above the arches, they might cause the falsework to settle and create cantilever action in the ribs, which would induce extraordinary stresses at their bases. He believed this a serious matter since the ribs were not as massive as those usually seen on deck arch structures on which the traditional decentering technique, with extra steel and concrete, was employed. McCullough and Gemeny, "Designing the First Freyssinet Arch to Be Built in the United States," 844–45. See also, Hadlow, "Oregon's Isaac Lee Patterson Bridge," 8–10.

[42]The concrete plant was on the north bank of the river. Workers there combined standard bags of cement with local aggregate and water in a "27-S Foote" non-tilting mixer. Then they elevated the mix by an 80-foot-high (later a 104-foot-high) wooden tower and distributed it through a wooden shoot to the forms. Concrete for piers, columns, and beams was designed for a 28-day compressive strength of 2,500 psi. For the deck slabs it was rated at 3,000 psi, and for the arch ribs, 5,000 psi. McCullough used the following ratios for the concrete of different compressive strengths: for 2,500 psi, 5 sacks of cement per cubic yard, with water-cement ratio of 0.80; for 3,000 psi concrete, 5.5 sacks of cement and a water-cement ratio of 0.75; and for 5,000 psi, 8.25 sacks of cement and a water-cement ratio of 0.60. G. S. Paxson and Marshall Dresser, "Concrete Arch Ribs of the Rogue River Bridge Decentered by Built-In Jacks," *Construction Methods* 15 (April 1933): 37–39.

[43]Gemeny and McCullough, *Application of Freyssinet Method of Concrete Arch Construction*, 54–58.

[44]Plowden, *Bridges: The Spans of North America*, 319. Interview, Robert W. Hadlow and Richard L. Koochagian with Ivan Merchant, Construction Inspector for OSHD at the Isaac Lee Patterson Memorial Bridge (1930-31), Gold Beach, Oregon, 16 August 1990, tapes held by ES, TS, ODOT.

[45]Van Leer, "Spanning the Mighty Rogue," 18.

[46]Ibid. The Oregon State Highway Department kept its own collection of newspaper clippings on highway department topics from 1916 to 1919. Beginning in 1932 it employed a Portland clipping service to assemble and file copies of articles from Oregon newspapers that covered any highway department topic. These were mounted on pages, with notation of publication date and the periodical's name. The OSHD maintained this professionally assembled clipping file from 1932 to 1950. The collection, 1916-19 and 1932-50, is located, in twenty-four records cartons as: Oregon State Highway Department Clipping File [CF-OSA], 78A-54, Highway Division Record, Oregon State Archives, Salem, Oregon. See Robert Withrow, "Throngs See Coast Road Link Opened," *Oregon Journal* (Portland), 29 May 1932, CF-OSA.

[47]Van Leer, "Spanning the Mighty Rogue," 18; Withrow, "Throngs See Coast Road Link Opened," "Rogue Bridge Name Honors Former Governor," Rogue Coast Supplement to the *Curry County Reporter* (Gold Beach, Oregon), 26 May 1982, 20. Initially, the planning committee had a difficult time finding monetary support for its dedication ceremony. The Oregon Coast Association promised $200 to cover expenses, but the OSHC denied former Oregon governor A. W. Norblad's request that it give $250 for the cause. Fiscal exigencies caused the commission to written April that it had "no legal authority" to expend funds for purposes of this kind, but it "fully supported" the enthusiasm for a dedicatory celebration. See Robert L. Withrow, to Leslie M. Scott, OSHC, 2 April 1932, A. W. Norblad to Scott, 31 March 1932, and Oregon State Highway Commission to Norblad, 14 April 1932, Folder 8-4, 76A-90/3, HDR, OSA. The OSHC truly dampened the local organizers' spirits. Secretary Withrow believed that the commission had its priorities wrong. He argued that the state of Oregon had a vested interest in seeing that the opening celebrations for the $600,000 bridge and the $17 million highway be a success. Completion of both projects, he believed, might breathe new life into once isolated fishing villages experiencing severe economic problems brought on by the Great Depression. Robert L. Withrow to C. B. McCullough, 23 April 1932, and 26 April 1932, Folder 8-4, 76A-90/3, HDR, OSA. The Patterson Bridge's cost of nearly $600,000 was part of the $2,575,520 spent on bridges along U.S. 101 up to May 1932. The latter sum was a portion of the $17 million expended on the entire highway. Withrow to McCullough, 6 May 1932, Folder 8-4, 76A-90/3, HDR, OSA; "Spanning over the mighty Rogue," 19. See also an untitled article by Withrow in *Oregon Journal Magazine* (Portland), 22 May 1932, CF-OSA, for a recapitulation of the events leading up to the completion of the Oregon Coast Highway and the Patterson Bridge. A program of events for the dedication is found in "Oregon Coast Highway Now Open for World Travel," *Coos Bay Times* (Marshfield), 27 May 1932, CF-OSA. As the twenty-eighth approached, donations poured in from private citizens for the celebration. The

local committee's financial worries vanished, see *Curry County Reporter* (Gold Beach, Oregon), 2 June 1932, CF-OSA.

[48]Many citizens believed that Herbert Hoover planned to ceremoniously open the bridge by telegraphic signal. See "President Hoover to Start Celebration," *Curry County Reporter* (Gold Beach, Oregon), 12 May 1932, CF-OSA. Many news accounts of the dedication ceremonies mentioned that in reality Hoover, and not Curtis, keyed a message to Gold Beach. While Curtis traveled to the White House on 28 May, he arrived there at 3:45 P.M. Eastern Time, one hour and fifteen minutes after the time that Hoover was to send the electrical impulse (11:30 A.M., Pacific Time). A record of the President's activities for the day shows that he was at the White House's Executive Office Building until he left for Camp Rapidan shortly after five o'clock in the afternoon. It did not mention any ceremony with a telegraph key. It is not known whether anyone in Washington, D.C., participated in the dedication of the Patterson Bridge and the Oregon Coast Highway. "The President's Day at the Executive Offices, 28 May 1932," *U.S. Daily* (Washington, D.C.), 31 May 1932. Concerning Hoover's deferring to Vice President Curtis, see Thomas T. Thalken, Director, Herbert Hoover Presidential Library, to Ray A. Allen, 12 September 1980, held by Allen, Portland, Oregon.

[49]"More than 1,000 Cars in Caravan for Celebration," *Coos Bay Times* (Marshfield), 30 May 1932; Withrow, "Throngs See Coast Road Link Opened"; and Withrow, "Great Throng Joins in Big Celebration," *Curry County Reporter* (Gold Beach), 2 June 1932 (quotes); "The Patterson Bridge," editorial, *Oregon Statesman* (Salem), 27 May 1932 (quote), CF-OSA.

[50]Albin LeRoy Gemeny continued with the BPR as a senior structural engineer. He died after suffering a heart attack on 1 December 1939. Other than his association with the Patterson bridge project, he was most noted for his work in the 1930s to revise highway bridge specifications in the United States. See "A. L. Gemeny Dies; Prominent Bridge Engineer," *Engineering News Record* 123 (7 December 1939): 738.

[51]*Biennial Report for 1923-24*, OSHC, 11; "Roy Klein Feted at Banquet Here," *Oregon Statesman* (Salem), 1 May 1932, CF-OSA; McCullough interview, 24.

Chapter 6

[1]"Steel Plaque is Presented for New Span," *Enterprise* (Oregon City), 28 November 1934 (quote).

[2]McCullough and Thayer, *Elastic Arch Bridges*, iii (quote).

[3]Ibid., 11-13.

[4]Ibid., 14-32.

[5]David P. Billington, "History and Esthetics in Concrete Arch Bridges," *Journal of the Structural Section, Proceedings of the American Society of Civil Engineers* 103, no. ST11 (November 1977): 2131, 2132; C. R. Young, review of *Elastic Arch Bridges* by Conde B. McCullough and Edward S. Thayer, in *Canadian Engineer* 31 May 1932, n.p.

[6]Conde B. McCullough, "Design of a Concrete Bowstring-Arch Bridge, Including Analysis of Theory," *Engineering News-Record* 107 (27 August 1931): 337-39. For

costs and construction schedules, see Job Record Cards for Wilson River Bridge (No. 1499), Big Creek Bridge (No. 1180), and Ten Mile Creek Bridge (No. 1181), BS, TS, ODOT.

[7]McCullough, "Design of a Concrete Bowstring-Arch Bridge, Including Analysis of Theory," 337-39; Robert W. Hadlow, "Wilson River Bridge at Tillamook, 1931, Tillamook County, HAER No. OR-39," 4-6, HAER-1990.

[8]McCullough, "Design of a Concrete Bowstring-Arch Bridge, Including Analysis of Theory," 337-39. Recent authors refer to McCullough's three bridges as the first tied arches in the United States. These claims are inaccurate. All-metal bowstring arches have existed since the mid- to late-nineteenth century. In addition, Marsh's bowstring arch bridge used this same concept in concrete and steel. But as suggested in the text, McCullough advanced the tied-arch form in concrete and steel reinforcing bar. The Washington State Highway Department first used a reinforced-concrete through tied arch in 1934 over the Duckabush River, in Jefferson County, on State Route No. 9. See Hadlow, "Wilson River Bridge," 7-9. See also "Duckabush River Bridge, State Route No. 9," Drawings, Bridge Design, Washington State Department of Transportation, Olympia, Washington, copies held by author.

[9]McCullough believed that a reinforced-concrete north viaduct was the best alternative because it would permit earlier use of the highway, present "a much more desirable appearance," "eliminate the uncertainty as regards the placement of such a high fill on movable sub-strata," eliminate the expense of continually adding more fill to one that would shrink over time, and, finally, they provided the best economic alternative for the site. Conde B. McCullough to Roy A. Klein, State Highway Engineer, 27 April 1931, copy in "Cape Creek Bridge File, No. 1113," ES, TS, ODOT. Cape Creek Bridge was constructed over a stream that empties into the Pacific Ocean between a large headland called "Devil's Elbow," and Heceta Head, a point named after the eighteenth century Spanish mariner and explorer Bruno de Hezeta who sailed near these shores in 1775. The region experienced little settlement by people of European descent until 1894, when a lighthouse and tender's residence were constructed on Heceta Head. McArthur, *Oregon Geographic Names*, 6th ed., 403; and Federal Writers' Program of the Work Projects Administration, *Oregon: End of the Trail*, 377. For costs, see "Oregon State Highway Commission Bridge Maintenance, Repairs and Renewals [1934]." Cape Creek Bridge (No. 1113), Maintenance Files, BS, TS, ODOT.

[10]"Engineering Antiquities Inventory for Cape Creek Bridge," TMs [photocopy], ODOT, 1982; Smith, Norman, and Dykman, *Historic Highway Bridges of Oregon*, 108; E. S. Hunter, Assistant State Highway Engineer, ODOT, to David Plowden, 14 June 1973, copy in "Cape Creek Bridge File, No. 1113," ES, TS, ODOT.

[11]"Highway of History," editorial, *Oregon Journal* (Portland), 28 May 1932, CF-OSA.

[12]"Report of State Highway Commission," *Biennial Report for 1931-32*, OSHC, 13-16.

[13]Seely, *Building the American Highway System*, 88-89. "U.S. Loan May Cover Coast Bridge Needs," *Register-Guard* (Eugene), 6 August 1932; "Coast Bridge Plan Backing Tendered," *News* (Eugene), 19 August 1932; "Coast Route Bridge Loan Endorsed," *Astorian-Budget* (Astoria), 9 August 1932; "Sawyer Has Coast Toll Plan That Will Not Increase State Debt," *Oregonian* (Portland), 25 August 1932, CF-OSA.

[14]Ernest W. Peterson, "McCullough Honored Guest at Banquet," *Oregon Journal* (Portland), 8 October 1935, CF-OSA.

[15]C. B. McCullough, "Timber Highway Bridges in Oregon," *Engineering News-Record* 109 (25 August 1932): 214-15.

[16]Ibid.; "Oregon Highway Department Plans Timber Bridges of Distinctive Type," *Oregonian* (Portland), 9 May 1932, CF-OSA.

[17]"U.S. Loan Urged for Coast Spans," *Oregonian* (Portland), 23 August 1932, CF-OSA.

[18]"Road Association Endorses 4 Toll Spans on Coast," *Coos Bay Times* (Marshfield), 29 August 1932. See "The Coast Bridges," editorial, *News* (Springfield), 1 September 1932 and "About Those Coast Bridges," editorial, *Astorian-Budget* (Astoria), 10 September 1932, CF-OSA, for examples of public opinion on the toll bridge proposal.

[19]See "Federal Funds Denied," *Oregonian* (Portland), 22 September 1932; and "Coast Bridges Can't Receive Federal Money," *Journal* (Newport), 28 September 1932, CF-OSA. Meanwhile, problems with the ferry service mounted. On 3 October 1932, the *Astorian-Budget* reported, in a United Press copyright story, that "Fifty occupants of 12 automobiles" on the beach at Newport awaited the "high tide to float the ferry, marooned on a sand spit." It had "broke down with engine trouble and drifted onto the sand bar before a pilot boat could come to its aid. Passengers stayed aboard for several hours in the hopes it would get off the bar, but finally were taken ashore by the coast guard [*sic*]." "Coast Group Still Seeks Five Bridges," *Astorian-Budget* (Astoria), 3 October 1932. Finally, inadequate markings on the vehicular approach to a ferry landing at Coos Bay caused a fatal automobile accident on 1 October 1932. See "Plainer Ferry Markings Asked," *Coos Bay Times* (Marshfield) 4 October 1932; and "State Ponders New Devices on Coast Highway," ibid., 14 October 1932, CF-OSA.

[20]"Road Session is Tomorrow," *Astoria-Budget* (Astoria), 5 April 1933; Harry N. Crain, "Federal Aid Funds to Provide Money to Complete Roads," *Capital Journal* (Salem), 1 May 1933. On Dolan's activities, see "Devers Says Sam Dolan Initiated Coast Bridges," *Gazette-Times* (Corvallis), 27 March 1934, CF-OSA.

[21]The 30-to-70 ratio was the standard split between the grant and loan portions of PWA financing packages. Harry N. Crain, "Liberalized Loan Conditions Boost Bridge Prospects," *Capital Journal* (Salem), 26 May 1933; "Coast Route Toll Bridges Get Approval," *Oregon Journal* (Portland), 1 June 1933, CF-OSA.

[22]"Toll Bridge Across the Umpqua River near Reedsport, Oreg.," *Senate Report 118*; "Bridge across Yaquina Bay, near Newport, Oreg.," *Senate Report 119*; "Toll Bridge Across Alsea Bay near Waldport, Oreg.," *Senate Report 120*; "Toll Bridge Across Coos Bay near North Bend Oreg.," *Senate Report 121*; "Bridge Across Siuslaw River near Florence, Oreg.," *Senate Report 122*, 73d Cong., 1st sess., Serial 9769; "3,400,000 Asked for Bridges," *Oregonian* (Portland), 26 May 1933; Harry N. Crain, "Approval of Loans for Coast Bridges Now Believed Sure," *Capital Journal* (Salem), 1 June 1933, CF-OSA.

[23]"Senate Approved McNary Toll Bridge Authorization," 7 June 1933, and "Bridge Engineers on Double Shift," *Capital Journal* (Salem), 20 June 1933 (quote), CF-OSA.

[24]"Five Bridge Designs to Be Rushed," *Oregon Journal* (Portland), 19 June 1933 (quote); "Senate Approved McNary Toll Bridge Authorization," 7 June 1933;

Sheldon F. Sackett, "Crow's Nest," editorial column, *Coos Bay Times* (Marshfield), 20 June 1933, CF-OSA.

[25]Merchant stayed with the OSHC bridge department until 1972, when he retired after 42 years with the department, the last 14 years as State Bridge Engineer, in which he followed in the footsteps of Glenn S. Paxson and P. M. Stephenson. Merchant died in Salem on 25 May 1997. In the spring of 1929, McCullough offered a senior seminar on elastic deformations of arch bridges, a topic he knew well. See the category "Civil Engineering," in *Schedule of Lectures, Recitations, and Laboratory Periods, Third Term 1928-29*, Oregon State Agricultural College, n.p. Karen Growth [Groth], "Conde B. McCullough, An Engineer with Soul," *Historic Preservation League of Oregon Newsletter*, no. 41, Summer 1986, 5 (quote).

[26]McCullough mentioned his hiring of designers with architectural training in C. B. McCullough to Aymar Embury, II, 6 May 1938, Office of General Files, ODOT. For Merchant quote see interview, Louis F. Pierce with Ivan Merchant, 4 June 1980, TMs, transcript held by Pierce, Junction City, Oregon. For Stephenson quote, see Stephenson interview, 11.

[27]"Plans for Two Coast Bridges are Complete, *Oregon Statesman* (Salem), 2 August 1933, CF-OSA. Harold Ickes, administrator of the PWA, announced on 22 June that the protocol for distributing loans and grants was not firmly in place. He asked states to hold off submitting public works proposals. He delayed again throughout the summer, raising tempers of representatives from several states. See "Devers Enroute Home after Presenting Loan Projects Data," *Capital Journal*, 23 June 1933, and Harry N. Crain, "Loans for Bridges to be Among First Belief of Devers," ibid., 1 July 1933; "State Boosts Bridge Request," ibid., 2 August 1933, CF-OSA.

[28]"McCullough Lauds Co-operation of Civil Port Units in Bridge Project," *Coos Bay Times* (Marshfield), 15 August 1933, CF-OSA.

[29]"Coast Bridge Plans Completed," *News* (Newport), 14 September 1933; "Soaring Prices of Materials Boost Bridge Estimates," *Capital Journal* (Salem), 8 September 1933. One speculates that President Roosevelt's inflationary policies of the early New Deal caused the rise in labor and material costs that prompted the revised funding request. "Bridge Building Anticipated Soon," *Oregon Statesman* (Salem), 9 September 1933; "Contracts on Coast Bridges Received Here," *Capital Journal* (Salem), 4 April 1934, CF-OSA.

[30]For newspaper publicity of the proposed bridges see "Sketches of Five Bridges to Span Major Ocean Inlets Along Oregon Coast Highway," *Oregon Journal* (Portland), 3 October 1933 and "Coast Route Spans Appear Near," *Oregonian* (Portland), 4 October 1933, CF-OSA. See Kenneth J. Guzowski, "Alsea Bay Bridge, Lincoln County, HAER No. OR-14," and "Yaquina Bay Bridge, Lincoln County, HAER No. OR-44," HAER-1990.

[31]"Coast Route Spans Appear Near," *Oregonian* (Portland), 4 October 1933, CF-OSA; Gary M. Link, "Siuslaw River Bridge, Lane County, HAER No. OR-10," 7-9, and "Umpqua River Bridge, Douglas County, HAER No. OR-45," 7-9, HAER-1990.

[32]Gary M. Link, "Coos Bay Bridge (McCullough Memorial Bridge), Coos County, HAER No. OR-46," 7-9, HAER-1990.

[33]Schwantes, *The Pacific Northwest*, 302.

[34]"Concerted Effort for Wooden Bridges Urged by Martin," *Register-Guard* (Eugene), 6 July 1933; "Want Bridge Built of Wood, Lumber Interests May Wreck Setup to Obtain Five Coast Spans if They Persist," *Harbor* (North Bend), 6 July 1933; "Lumbermen to Meet to Protest Concrete for 5 Coast Bridges," *Sentinel* (Cottage Grove), 7 July 1933 (quote), CF-OSA. The *Sentinel* also reported that McCullough, in talking to Crow, arrogantly referred to the lumber industry as the "lousy lumber industry" and that the coast highway bridges would be of steel and concrete or not built. "'Lousy' Lumber Seems Suitable for California," editorial, *Capital Press* (Salem), 7 July 1933 (quote), CF-OSA.

[35]"Highway Session Hot On Bridges," *Oregonian* (Portland), 12 July 1933 (quote); Harry N. Crain, "Selfish Interest Bared in Protest over Bridge Plan," 12 July 1933; and "Backers of Coast Road Bridges Saw Wood Not Wanted," *Capital Journal* (Salem), 8 July 1933; "Road Unit Seeks to Push Project," *Coos Bay Times* (Marshfield), 7 July 1933, CF-OSA.

[36]"The Five Coast Bridges," editorial, *Oregonian* (Portland), 5 October 1933. See also "Bridges and Red Tape," editorial, *Oregonian* 13 October 1933, CF-OSA.

[37]"Bridge Projects Being Speeded Up by Oregon PWA," *Coos Bay Times* (Marshfield), 9 October 1933. The war of words escalated when former Oregon governor Oswald West, early promoter of the Columbia Gorge Highway and champion of the coast highway bridge project, went head-to-head with Bert E. Haney, chairman of the Oregon State PWA advisory board. West accused Haney, a Democratic candidate for governor, of stalling the project for political gains. He argued, "No man . . . should be permitted to run for governor on the flattened stomachs of the hungry." Haney, he believed, "Sees in the distribution of the [PWA] funds an opportunity to play politics." Haney, though, denied all of West's "baseless" charges. See "Oswald West Fears 4 Oregon Bridges will be Lost through Pigheadedness," *Oregonian* (Portland), 12 October 1933, CF-OSA.

[38]"Plans for 5 Bridges in Hands of PWA," *Capital Journal* (Salem), 31 October 1933; "PWA Funds Granted for Five Bridges," *Oregon Journal* (Portland), 7 January 1934. For comment against for tolls see "Tolls Unfavored by Road Group," *Coos Bay Times* (Marshfield), 12 June 1933 "Make Coast Bridges Free," *Oregonian* (Portland), 13 December 1934 and "No Tolls for Coast Bridges," editorial, *Register-Guard* (Eugene), 24 January 1934; and "Bridges Should be Free," *Capital Press* (Salem), 12 January 1934. For MacDonald's comments on tolls, see "Speed Plans for 5 Bridges Along Coast," *Capital Journal* (Salem), 19 June 1933; "Another Hurdle Taken," editorial, *Astorian-Budget* (Astoria), 5 December 1933. For comment favoring tolls see "The Coast Bridges," editorial, *News* (Springfield), 1 September 1932 and "Bend Demands Votes Against Toll Measure," *Capital Journal* (Salem), 8 February 1935. In editorializing on toll bridges versus ferries, the Oregon City *Enterprise* reported that there was nothing wrong with the slow and unreliable ferries. "Only those who insist on burning up the roads register impatience at the slight delays. The sensible traveler finds helpful respite in the enforced stops and the occasional traveler a bit of novelty in the ferries." "Toll Bridges for Ferries," 30 August 1932, CF-OSA.

[39]OSHC, *Biennial Report for 1935-36*, 16-17; OSHC, *Biennial Report for 1933-34*, 17-18. See also, "Board Plans Free Tolls for Bridges," *Oregon Journal* (Portland), 21 February 1935. The *Coos Bay Times* issued probably the most convincing series of

editorials making the case for free bridges; see "The Birth of the Coast Bridges," 10 January 1935; "Free Bridges Bill All-Important," 11 January 1935; "General Revenue Would Pay Cost," 12 January 1935; "Tolls Menace To Gain in Travel," 14 January 1935; "'Break Faith' Charge is Error," 15 January 1935; "Free Spans Win Staunch Support," 16 January 1935; and "Oregon Cannot Afford Tolls," 17 January 1935, CF-OSA.

[40]For the text of Kerr's conferring speech see "Mc—Degrees, Honorary" File, Oregon State University Archives, Corvallis. See also press coverage of McCullough's degree conference in "C. B. McCullough Honored by Degree from State College," *Journal of Commerce* (Portland), 7 June 1934, CF-OSA.

[41]Smith, Norman, and Dykman, *Historic Highway Bridges of Oregon*, 109.

[42]"Steel Bridge Awarded Prize is of Three Hinged Tied Arch Type," *Engineering News-Record* 113 (23 August 1934): 248.

[43]Ibid.

[44]"Steel Plaque is Presented for New Span," *Enterprise* (Oregon City), 28 November 1934 (quotes). Actually, McCullough was not present for the award ceremony because of illness. See "McLoughlin Span Lauded as Plaque for Beauty Awarded; Scott Accepts," *Oregonian* (Portland), 28 November 1934, CF-OSA. Mervyn Stephenson recalled that McCullough submitted to the American Institute of Steel Construction's competition a watercolored sketch of the bridge. On it he tinted the steel arches green. In reality, the OSHD covered the exposed metal portions of this bridge, and all others, with inexpensive black lead paint. When the McLoughlin Bridge won the competition, McCullough sent out crews to repaint the metal arches green to match the watercolor. From then on, for decades, the OSHD painted metal portions of its bridges green—now known as "ODOT Green" in some color chip charts for bridge coatings. Stephenson interview, 16.

[45]"Bridge Program Hailed as Major Step in Progress," *Coos Bay Times* (Marshfield), 17 January 1934, CF-OSA.

[46]"$685,040 Low Offer on First Coast Bridge," 5 April 1934 and "Bids on Last Coast Bridge to be Opened at Meeting on June 7," *Capital Journal* (Salem), 24 May 1934; "Start to be Made on Coast Bridges," *Oregon Statesman* (Salem), 27 July 1934. For the increases in cost estimates, see "PWA to Allot Extra Costs of Coast Bridges," *Capital Journal* (Salem), 10 May 1934, CF-OSA.

[47]Guzowski, "Yaquina Bay Bridge," HAER-1990. Thomas H. MacDonald was a great supporter of highway beautification. As of 1 January 1934, the federal government, through the Bureau of Public Roads, required that at least 1 percent of all federal funds allocated for highway projects be devoted to roadside beautification. Because of this, wayside parks became an integral part of Oregon's coast bridge project. The Works Progress Administration oversaw a portion of Oregon's roadside beautification program through relief employment. See "To Beautify Highways," editorial, *Oregon Journal* (Portland), 4 November 1933, CF-OSA. See also, "WPA Projects," and "Roadside Beautification," *Biennial Report for 1935-36*, OSHC, 47-48 and 69-70.

[48]C. B. McCullough, "Self Liquidating Plan for Oregon's Coast Highway," *Engineering News-Record* 114 (6 June 1935): 814-15; C. B. McCullough, "Remarkable Series of Bridges on Oregon Coast Highway," *Engineering News-Record* 115 (14 November 1935): 679.

[49]Dexter Smith created the reinforced-concrete deck arch approaches for the Yaquina Bay Bridge. They were also used on the Alsea Bay Bridge and the Coos Bay Bridge. Ivan Merchant was only twenty-seven years old when he designed the Yaquina Bay Bridge's central arches. "Yaquina Bay Bridge," *Oregon Motorist*, May 1936, 11; Guzowski, "Yaquina Bay Bridge," HAER-1990; Elizabeth Shellin Atly, "C. B. McCullough and the Oregon Coastal Bridges Project," TMs, 1977, pp. 12-14, copy held by author; *Biennial Report for 1935-36*, OSHC, 59.

[50]For a complete discussion of the bridge's features, see O. A. [*sic*]. Chase, "Design of Coast Highway Bridges," *Civil Engineering* 6 (October 1936): 647-51. See also "Alsea Bay Bridge," *Oregon Motorist*, May 1936, 10; and Atly, "C. B. McCullough and the Oregon Coastal Bridges Project," 12-14; *Biennial Report for 1935-36*, OSHC, 59.

[51]Link, "Siuslaw River Bridge," 7-9, HAER-1990; Atly, "C. B. McCullough and the Oregon Coastal Bridges Project," 16-17. McCullough previously co-authored a long U.S. Bureau of Public Roads report on the mechanical portion of moveable bridges. With Albin Gemeny and W. R. Wickerham, he wrote *Electrical Equipment on Movable Bridges*, Technical Bulletin No. 265, for the BPR in 1931; OSHC, *Biennial Report for 1935-36*, 59.

[52]Link, "Umpqua River Bridge," 7-9, HAER-1990; Atly, "C. B. McCullough and the Oregon Coastal Bridges Project," 19-21; "Umpqua River Bridge is Third Costliest of Five Coast Structures," *Courier* (Reedsport), 25 September 1936, CF-OSA. *Biennial Report for 1935-36*, OSHC, 59.

[53]Total length of the approach spans, each of a different dimension, was 2,700 feet. The cantilevered section included 458-foot approach sections flanking a midsection of 793 feet. Link, "Coos Bay Bridge (McCullough Memorial Bridge)," 7-10, HAER-1990; "Coos Bay Bridge," *Oregon Motorist*, May 1936, 7. McCullough and the OSHC proposed including parks near the ends of each bridge not only for recreational use of motorists and pedestrians, but also to curb private business encroachment. They hoped that the parks might act as buffers by preventing entrepreneurs from cluttering the landscape near the bridges with structures that detracted from both the spans' picturesqueness and the scenic attraction of the bays and estuaries. See "McCullough Tells Rotary Club of Coast Bridges," *Gazette-Times* (Corvallis), 23 November 1933, CF-OSA. McCullough explained his attention to detail on the piers in "Remarkable Series of Bridges on Oregon Coast Highway," 679. *Biennial Report for 1935-36*, OSHC, 59.

[54]G. S. Paxson to Charles E. Wilson, *Coos Bay Times*, 2 October 1941, Miscellaneous Records, BS, TS, ODOT.

[55]Nathan Douthit, *A Guide to Oregon South Coast History* (Coos Bay: River West Books, 1986), n.p.

[56]Plowden, *Bridges: The Spans of North America*, 255.

[57]"Tourist Travel Stimulation," *Biennial Report for 1935-36*, OSHC, 72-73.

[58]All bridges opened for traffic a few weeks or months before actual completion. See *Biennial Report for 1935-36*, OSHC, 54-59. The dignitaries included John Williams, Chief of the Siletz Indians. On the dedications, see "Dedication of Bridge Occasion in Building Highway," *Capital Journal* (Salem), 3 October 1936, and "Newport Pays Honor to Last Coast Bridge," *Oregon Journal* (Portland), 4 October 1936, and *News* (Newport), 8 October 1936, CF-OSA.

[59]Plowden, *Bridges: The Spans of North America*, 319 (quote); Christopher Boehme, "The Oregon Coast Bridges," *Pacific Northwest*, July 1988, 22 (quote).

[60]C. B. McCullough to J. E. Mackie, 7 September 1937, Office of General Files, ODOT.

Chapter 7

[1]"Conde B. McCullough—Bridges," editorial, *Register-Guard* (Eugene), 7 May 1946.

[2]Ernest W. Peterson, "1265 Miles of Inter-American Highway Built," *Oregon Journal* (Portland), 28 March 1937, CF-OSA; *America's Highways*, FWHA, 139-40, 522. See also, "Proposed Inter-American Highway—Report Prepared by the Department of Agriculture Transmitted to the Secretary of State," *S. Doc. 224*, 73d Cong, 2d sess., 13 passim.

[3]Peterson, "1265 Miles of Inter-American Highway Built."

[4]*America's Highways*, FWHA, 522-24; *Annual Report for 1938*, Bureau of Public Roads [BPR], Department of Agriculture, 62-64; George R. Stewart, *N.A. 1—The North-South Continental Highway—Looking South* (Boston: Houghton Mifflin Co., and Cambridge, MA: The Riverside Press, 1957), 26.

[5]*Annual Report for 1936*, BPR, 61; Stewart, 27; *Annual Report for 1938*, BPR, 62-64. See also *America's Highways*, FWHA, 522-25; "McCullough Leaves Soon for Costa Rica to Build," *Oregon Statesman* (Salem), 6 October 1935.

[6]"McCullough Honor Guest at Banquet," *Oregon Journal* (Portland), 8 October 1935, CF-OSA.

[7]"McCullough Gets Plaque at Banquet," *Oregon Journal* (Portland), 9 October 1935; "C. B. McCullough," editorial, *Capital Journal* (Salem), 9 October 1935; "McCullough Paid Honors by Staff," *Oregon Statesman* (Salem), 11 October 1935, CF-OSA.

[8]"Engineer Skelton Given Promotion," *News-Review* (Roseburg), 6 November 1935, CF-OSA; Stephenson interview, 4.

[9]"Dexter Smith to Succeed Bridge Engineer on Bay," *Coos Bay Times* (Marshfield), 13 November 1935; "Archibald Said In Line for Job," *Coos Bay Times* (Marshfield), 8 November 1935, CF-OSA.

[10]"Bridge Construction Along the Inter-American Highway," *Bulletin of the Pan American Union* 70 (1936): 63-64; Conde B. McCullough, "Bridging the Rio Chiriqui on the Pan-American Highway," *Engineering News-Record* 117 (26 November 1936): 757-58; Conde B. McCullough and Raymond Archibald, "Bridging the Rio Choluteca with a Two-Span Suspension Structure," *Engineering News-Record* 118 (21 January 1937): 87.

[11]Ibid.; McCullough, "Bridging the Rio Chiriqui," 758 (quote).

[12]McCullough, "Bridging the Rio Chiriqui," 757-58.

[13]This bridge and the other two were designed with an "H-15" live load rating. McCullough, "Bridging the Rio Chiriqui," 757-58.

[14]McCullough and Archibald, "Bridging the Rio Choluteca," 87-88.

[15]Tufa stone was porous rock formed as a calcareous deposit from springs or streams. McCullough and Archibald, "Bridging the Rio Choluteca," 87-88.

[16]The towers were 58 feet tall. McCullough and Archibald, "Bridging the Rio Choluteca," 87-88.

[17]Conde B. McCullough and Raymond Archibald, "Self-Anchored Eyebar Cable Bridge," *Engineering News-Record* 118 (17 June 1937): 912-13.

[18]Eyebars were bars of steel with holes drilled in their ends. Pins connected a number of eyebars in series much like a chain. McCullough used them instead of wire cable as much for ease of transport to the construction site as he did for architectural variety. McCullough and Archibald, "Self-Anchored Eyebar Cable Bridge," 912-13; Condit, *American Building*, 234.

[19]McCullough and Archibald, "Self-Anchored Eyebar Cable Bridge," 913 (quotes).

[20]*Annual Report for 1938*, BPR, 63; Stewart, *N.A. 1—The North-South Continental Highway—Looking South*, 27; McCullough and Archibald, "Bridging the Rio Chiriqui," 87-88. Ninety percent of the $1 million congressional appropriation was used on the bridges by the end of fiscal year 1938. Of this, over 70 percent was for materials and about 20 percent for engineering services and personnel. The Central American republics contributed an additional $710,000 to the Inter-American Highway bridges project. *Annual Report for 1938*, BPR, 63.

[21]"Bridge Engineer Back from Central America, *News-Telegram* (Portland), 1 February 1937; C. B. McCullough for United Press, "McCullough and Archibald Send Their Greetings," *Coos Bay Times* (Marshfield), 1 June 1936 (quote), CF-OSA. Raymond Archibald did not return to the OSHD when the project was completed. Instead, he elected to remain with the BPR as a structural engineer. Stephenson interview, 4-5.

[22]"McCullough Talk Set for Tuesday," 7 February 1937, and "McCullough Talks of Maya Culture," *Oregon Statesman* (Salem), 10 February 1937, CF-OSA. The Salem Arts Council sponsored McCullough's lecture on 9 February that attracted a standing-room only audience. McCullough gave similar lectures to other groups including the Salem and McMinnville chambers of commerce. See "M'Cullough Tells Chamber of Guatemala," *Capital Journal* (Salem), 22 February 1937; and "McColloch [sic] Tells Chamber of South America [sic]," *News-Reporter* (McMinnville), 25 March 1937, CF-OSA. Some participants in the Marshfield celebration called McCullough "Papa Dionne," after the recently arrived Canadian-born Dionne quintuplets. For a report of the Marshfield dinner, see "Designer of Coast Spans Given Laud: C. B. McCullough Comes Back to View His Work," *Coos Bay Times* (Marshfield), 20 February 1937, CF-OSA.

[23]"Highway Department Organization," *Biennial Report for 1937-38*, OSHC, 28-32. See especially pp. 28-30.

[24]McCullough interview, 25-26.

[25]Stephenson interview, 5. McCullough's son recalled that even though McCullough stood up for his principle when he believed it correct, he also admitted he was wrong if he were convinced of it. McCullough interview, 25-26 and 28; Merchant interview, 11 (quote).

[26]See McCullough, Conde B. *Determination of Highway System Solvencies.* Technical Bulletin No. 8. Salem: OSHD, 1937; Baldock, R. H. and C. B. McCullough. *The Merit System for Engineering Personnel.* Technical Bulletin No. 9. Salem: OSHD, 1938; and McCullough, C. B., John Beakey, and Paul Van Scoy. *An Analysis of the Highway Tax Structure in Oregon.* Technical Bulletin No. 10. Salem: OSHD, 1938.

[27]Conde B. McCullough, Glenn S. Paxson, and Dexter R. Smith, *An Economic Analysis of Short-span Suspension Bridges for Modern Highway Loadings,* Technical Bulletin No. 11 (Salem: OSHD, 1938), 1-2. See also C. B. McCullough, G. S. Paxson, and

Dexter R. Smith, *Rational Design Methods for Short-span Suspension Bridges for Modern Highway Loadings,* Technical Bulletin No. 13 (Salem: OSHD, 1940), 1–3 passim. McCullough also was a discussion participant in an American Society of Civil Engineers' meeting by correspondence on the merits of eyebar-type suspension bridges. See C. H. Gronquist, "Simplified Theory of the Self-Anchored Suspension Bridge," with discussion by A. J. Meehan, C. B. McCullough, Jaroslav J. Polivka, William Bertwell, A. A. Eremin, and C. H. Gronquist, Paper No. 2151 in American Society of Civil Engineers, *Transactions* 107 (1942): 979–83. Dexter Smith left the OSHD briefly in the 1940s to design for the Washington Toll Bridge Authority a suspension structure to replace the 1940 Tacoma Narrows suspension bridge ("Galloping Gertie") that collapsed due to excessive dynamic forces. Charles E. Andrew, *Final Report on Tacoma Narrows Bridge, Tacoma, Washington* ([Olympia]: [Washington Toll Bridge Authority], 1952), 7, 18–27.

[28]McCullough, Paxson, and Smith, *An Economic Analysis of Short-span Suspension Bridges for Modern Highway Loadings,* 83–84; C. B. McCullough, Glenn S. Paxson, and Dexter R. Smith, *The Derivation of Design Constraints for Suspension Bridge Analysis (Fourier-series Method),* Technical Bulletin No. 14 (Salem: OSHD, 1940), 1–2, 239–47; C. B. McCullough, Glenn S. Paxson, and Richard Rosencrans, *The Experimental Verification of Theory for Suspension Bridge Analysis (Fourier-series Method),* Technical Bulletin No. 15 (Salem: OSHD, 1942), 1–3.

[29]McCullough, Paxson, and Smith, *An Economic Analysis of Short-span Suspension Bridges for Modern Highway Loadings,* 83, 84 (quotes); C. B. McCullough, G. S. Paxson, and Richard Rosencrans, *Multiple-Span Suspension Bridges: Development and Experimental Verification of Theory,* Technical Bulletin No. 18 (Salem: OSHD, 1944), 1–3, 111–12, 1 (quote).

[30]Washington Department of Highways and the Oregon Department of Transportation, *Report on Trans-Columbia River Interstate Bridge Studies,* Technical Bulletin No. 16 (Salem: Oregon State Highway Commission, 1944), Division IV, 1–3.

[31] See Friedrich Bleich, Conde B. McCullough, Richard Rosencrans, and George S. Vincent, *The Mathematical Theory of Vibration in Suspension Bridges,* Bureau of Public Roads, Department of Commerce (Washington DC: Government Printing Office, 1950), 1ff.

[32]McCullough interview, 6–7.

[33]Interestingly, the McCulloughs were confirmed alongside Glenn S. Paxson and his wife. Also, many other OSHD employees in Salem were members of St. Paul's Episcopal Church, including the Robert Baldock family. "Confirmations" in *Canonical Church Register,* St. Paul's Episcopal Church, Salem, Oregon, for 3 November 1940, p. 232.

[34]Smith, Norman, and Dykman, *Historic Highway Bridges of Oregon,* 128–129, 293–294.

[35]McCullough interview, 26–27.

[36]John Roddan "Jack" McCullough became a member of the U.S. Naval Reserve in late 1941, eventually receiving a commission. He served two years on the destroyer *U.S.S. Overton,* in the South Pacific. McCullough and his wife each wrote to Jack every day during the war. A small portion of this collection remains and reveals McCullough's dedication to his son and concern for his welfare. In one letter, McCullough wrote to Jack about being committed to a cause, that of the military,

especially during time of war. McCullough had been a captain of a U.S. Army engineering corps at the Oregon Agricultural College during WWI. In WWII, he was a major attached to Brigadier-General Ralph P. Cowgill's Oregon State Guard Commander's Office. So military organization was not something foreign to him. He told Jack that his service to his country was his sacrifice, "that for a while you and all of the rest of the young men who are serving their country in the same capacity must suppress individuality, both in desire and action." C. B. McCullough to Jack, 30 January 1942 (quotes here and in text); C. B. McCullough to John W. Pollock, 10 August 1945; and C. B. McCullough to Jack, 10 July 1942 (quotes), held by Mrs. John R. McCullough and John B. McCullough, Salem, Oregon [JRM]; McCullough interview (quotes).

[37]James T. Brand, "Foreword," in McCullough and McCullough, *The Engineer at Law*, 7-10, 7 (quote). See also Joseph. M. Devers' "Forward," 11-13. William Tugman, editor of the Eugene *Register-Guard* believed that McCullough wrote the book as much to keep his mind off his son and the war. "Conde B. McCullough," editorial, *Register-Guard* (Eugene), 7 May 1946, CF-OSA. John R. McCullough told Louis Pierce that he wrote little, if anything, for the book. His father asked him to review manuscripts of a few chapters, but he did not actually write any of it. McCullough interview, 4-5.

[38]Salem Long Range Planning Commission, *A Long Range Plan for Salem, Oregon: First Annual Progress Report* (Salem, 1947), 1 (quote).

[39]Ibid., 72, 74.

[40]"Famous Highway Bridge Engineer Dies at Salem, *Coos Bay Times* (Coos Bay [Marshfield until 1944]), 7 May 1946, CF-OSA, mentioned that McCullough planned to leave Salem on Wednesday, 8 May for Central America. See also "Engineer Law Get Review," *Journal of Commerce* (Portland), 29 May 1946.

[41]Immediate cause of death was listed as cerebral hemorrhage with hemiplegia of right side, due to hypertension. "Standard Certificate of Death of Conde B. McCullough," Date of Death, 6 May 1946, State File No. 3553, held by Oregon Health Division, Center for Health Statistics, Portland. Probate records from McCullough's estate contained a number of final statements from local thirty-day charge accounts, including two from pharmacies. Those from Quisenberry's Central Pharmacy and South Salem Pharmacy showed a number of entries during his last month for nitroglycerine capsules to treat angina and aminophylline to treat bronchial spasms. McCullough had endured a severe bout of bronchitis earlier in 1946. His health at the time of his death appeared good but not exceptional. "C. B. McCullough," Case No. 12526, Probate Records, Marion County, Oregon. For newspaper obituaries, see CF-OSA for early May 1946. Periodicals in western Oregon carried multi-columned death notices for McCullough. Those in other parts of the state ran portions of wire service obituaries. All of them except for the Portland *Oregonian* erroneously reported his cause of death as a heart attack. This newspaper correctly reported it as a stroke. For a representative obituary, see "Famed Oregon Bridge Engineer Dies Suddenly, *Capital Journal* (Salem), 6 May 1946. A death notice also ran in the *New York Times*. See also an American Society of Civil Engineers' "Memoir" to McCullough in Glenn S. Paxson, "Conde Balcom McCullough, M. ASCE," in *Transactions of the American Society of Civil Engineers* 112 (1947): 1489-91.

[42]Honorary pallbearers included Oregon Supreme Court Justices Arthur Hay, James T. Brand, Hall Lusk, and Chief Justice Harry Belt; Eugene *Register-Guard* editor William Tugman; past and present highway commissioners, including T. H. Banfield, Arthur W. Schaupp, and Merle R. Chessman; and Oregon Governor Charles Sprague. Highway department employees included Glenn S. Paxson, Ellsworth G. Ricketts, Albert G. Skelton, Orrin A. Chase, and P. M. Stephenson, who were all members of McCullough's bridge department since its beginning. McCullough was cremated and his ashes placed in an urn at Mount Crest Abbey Mausoleum, in Salem's City View Cemetery. "Engineers Are Pallbearers At Rites Today," 8 May 1946, and "Many Attend Services for Noted Engineer," 9 May 1946, in *Oregon Statesman* (Salem), CF-OSA.

[43]"Hail and Farewell," editorial, *Oregon Statesman* (Salem), 7 May 1946; "Conde B. McCullough," editorial, *Capital Journal* (Salem), 6 May 1946; "Conde B. McCullough—Bridges," editorial, *Register-Guard* (Eugene), 7 May 1946; Don Upjohn, "Sips for Supper," column, *Capital Journal* (Salem), 6 May 1946, quotes, CF-OSA. William W. Stiffler, assistant maintenance engineer, and W. C. Williams, a division engineer, both replaced McCullough as Baldock's assistant engineers. See McCullough interview, 15.

[44]See "Resolution in appreciation of service of Conde Balcom McCullough to the State of Oregon by the Oregon State Highway Commission," 14 May 1947, copy in File 159a, Box 2, Correspondence of Chief Counsel—J. M. Devers, Accession No. 68A-34-1, Highway Division Records, OSA. For Banfield's dedicatory speech see "Address to be Presented by Mr. T. H. Banfield at the Ceremonies Incident to the Dedication of the Coos Bay Bridge in honor of the late Mr. C. B. McCullough, August 27, 1947," in "C. B. McCullough File," Environmental Services, Oregon Department of Transportation, Salem. Henry Petroski, *Engineers of Dreams: Great Builders and the Spanning of America*, (New York: A. A. Knopf, 1995), 344.

Chapter 8

[1]Condit, *American Building*, 251.

Epilogue

[1]Andy Booz, "Profile: Phil Rabb—Blending Form, Function, Public Ideas," *VIA* (Oregon Department of Transportation), September 1990; "Alsea Bay Bridge, Waldport, Oregon, Oregon Coast Highway (U.S. 101), Final Environmental Impact Statement," (Salem: Oregon Department of Transportation and the Federal Highway Administration, 1986), 2F-5F. HNTB was also a descendant firm of Waddell and Harrington, later in part, Harrington, Howard, and Ash.

[2]Carmel Finley, "Seattle Company makes Low Bid on Alsea Bridge," *Portland Oregonian*, 13 May 1988; Hasso Hering, "Alsea Bridge Nearing Completion," *Albany, Oregon, Democrat-Herald*, 29 June 1991; "Old, New Spans Revered," *VIA* (Oregon Department of Transportation), October 1991; Charlotte Snow,

"Demolition Sinks Landmark Bridge at Alsea," *Salem, Oregon, Statesman-Journal*, 2 October 1991.

[3]Tami Barnes, "Portions of Old Alsea Bay Bridge Find Home," *Newport, Oregon, News-Times*, 12 December 1991.

[4]For addition information on the Oregon Department of Transportation's bridge preservation program, see Frank J. Nelson, "The Oregon Department of Transportation Program for Restoring and Preserving its Historic Bridges," Conference Proceedings, "Preserving the Historic Road in American: The Second National Conference on Historic Roads," Morristown, NJ, 6-9 April 2000.

[5]See Eric DeLony, *Landmark American Bridges* (New York: Bulfinch Press and the American Society of Civil Engineers, 1993): 125 (quotes), 127-35. DeLony was so impressed by McCullough's work that he used images of the Coos Bay Bridge for the volume's dust jacket.

[6]See "Top People of the Past 125 Years," *ENR*, 30 August 1999, 27 (quote), 47-48.

Glossary

Words in italics are defined elsewhere in the glossary.

abutment—The support at each end of a bridge where the *roadway* meets the ground. An abutment may include a retaining wall to hold dirt in place.

anchorage—Cables are the main supports for *suspension bridges*. An anchorage holds the cables in place, and attaches them to the ground.

approach—The structure that carries traffic from land onto the main bridge.

arch bridge—One of three main bridge types, along with *beam* and *suspension* bridges. An arch is a curved structure that acts in *compression*—forces pushing together in the arch hold the bridge up. Many of McCullough's bridges are of this type.

balustrade—A decorative railing, usually constructed of concrete or stone. The individual posts are called balusters.

bascule bridge—One of three main movable bridge types, along with *vertical lift* and *swing* bridges. "Bascule" is a French word meaning seesaw. A bascule bridge allows the *roadway* to tilt up out of the way of river traffic using *counterweights* as counterbalance. Bascule bridges are either *single-* or *double-leaf.*

bascule pit—The space inside a bascule bridge housing the *counterweight*. Not all bascule bridges have bascule pits.

beam bridge—One of three main bridge types, along with *arch* and *suspension* bridges. A beam is a type of structure that carries loads in *bending*. Possibly the first beam bridge was a downed tree, placed across a stream or canyon so that people could walk across. Large beams spanning between main supports are called *girders*. One variation of a beam bridge is the *truss bridge.*

bearing—A device for transmitting the *load* from a bridge to its supports (see *abutment, bent,* and *pier*). Bearings are designed to allow for the movements of the bridge relative to its supports caused by *bending* and temperature changes.

bending—*Loads* on a *beam bridge* produce bending, which is caused by a combination of *compression forces* in one side of the beam and *tension* forces in the opposite side. By definition, all beams bend; if it does not bend, it is not a beam.

bent—An upright tower on land that supports a bridge. An upright tower in water that supports a bridge is called a *pier*. Bents and piers are parts of a bridge's *substructure*. (The words *bent* and *bending* here are not related.)

box girder—A steel or concrete *girder* with a hollow cross section, designed to give strength without excessive weight.

bridge tender/operator—A person employed to open, close, and help maintain movable bridges.

This glossary is excerpted from *The Portland Bridge Book*, 2nd edition, revised and expanded (Oregon Historical Society Press, 2001), and was written by Sharon Wood Wortman and Ed Wortman. It appears with permission of the book's author, Sharon Wood Wortman.

brittle—Instead of bending or stretching when overloaded, brittle materials (such as glass, cast iron, and concrete without steel reinforcing bars) break abruptly. Brittle materials pose dangers because they give little warning of overloading before breaking.

cable-stayed bridge—A variation on the conventional *suspension* bridge. The *roadway deck* is supported by a series of straight cables attached to towers and sloping down to the deck.

caisson—A type of *foundation* for a bridge or other structure, which may be used on land or in water. In general, caissons are open-bottom boxes or cylinders that are sunk into the ground to hold back the dirt and rock around the sides of a hole that it is filled with concrete or stone to support a structure. Different types of caissons are used, varying from cylinders a few feet in diameter up to boxes 100 feet or more in width. They can be made of concrete, steel, or wood. Caissons are installed by digging out the dirt and rock from inside, then pushing the caisson into the ground by adding weight to it. See also *coffer dam, crib*.

camber—An upward curve built into a bridge to anticipate *deflection* that will occur when *loads* are applied to the structure.

cantilever bridge—A type of *beam* bridge that came into general use for long spans during the railway era, when engineers found that bridges more rigid than *suspension* bridges (as they were then built) were needed to carry heavy trains and freight. A cantilever is a bracket or arm that juts out (like a diving board). A cantilever bridge is made up of two such brackets or arms, anchored at one end and free-hanging at the other. Many bridges are built using the cantilever principle in order to save materials and simplify construction. Cantilever bridges typically have two cantilever sections reaching out from each end, plus a suspended section joining the two cantilever sections in mid-span. The weight of the cantilevered sections is balanced by anchor *spans* on the backside of the *piers*. The Conde B. McCullough Memorial Bridge at Coos Bay is a good example of a cantilever bridge. Although not usually recognized as such, *double-leaf bascule* bridges are a form of cantilever bridge where the weight of the cantilevered sections is balanced by *counterweights*.

catenary—The natural curve formed by a cable draped from its ends. The main cables of a *suspension* bridge hang from the bridge towers in roughly this shape. (Once the *deck* is hung from the cable, the shape changes from a catenary to more of a *parabolic*-shaped curve.)

cathodic protection—An innovative method to prevent damage to reinforced concrete structures from salty ocean air and water, which can penetrate concrete and rust the reinforcing steel, by forcing corrosion of a sacrificial metal instead. The necessary connection between the higher energy, more active sacrificial metal and the reinforcing steel is made by means of direct current in the Oregon Department of Transportation's bridge preservation program. An arc-sprayed layer of solid zinc is placed on the exterior of the bridge, exactly following the rich artistic detailing of the structure. The zinc layer is then connected to the reinforcing steel through several heavy-duty power supplies that automatically adjust the current to ensure the exterior zinc and not the reinforcing steel corrodes. If the sacrificial zinc layer is replaced every twenty-five to thirty years, the bridges can indefinitely survive ocean environments.

chord—The main top or bottom structural member of a *truss bridge*.

clear span—Horizontal distance between two adjoining supports.

cofferdam—A temporary watertight enclosure built into a riverbed to permit construction of dams, bridge foundations or other structures. It normally is built in place by driving *piling* in a rectangular or circular pattern. The water inside the cofferdam is then pumped out, so that construction of the structure can proceed in open air. Cofferdams are similar to *caissons*. The primary difference is that cofferdams are removed when the structure is finished whereas caissons become a permanent part of the structure.

compression/compressive force—A force that tends to make something shorter, or pushes or squeezes it together. *Arch* bridges are in compression, as are portions of *beam* and *truss* bridges.

continuous-span bridge—A bridge that extends as one piece over multiple supports (*piers* and/or *bents*) without gaps or hinges, so that the *stresses* of the bridge are distributed over the entire structure. Continuous-span bridges can be built with *girders* or *trusses*, and use less material than a structure made up of a series of independent *spans*.

counterweights—*Vertical lift* and *bascule* bridges require counterweights to balance the weight of their movable *spans*. Counterweights usually are made of concrete.

covered bridge—A *beam* or *truss* bridge with the roadway protected by a roof and side walls, usually made of wood. This protects the roadway from rain and snow so the bridge will last longer. McCullough's Grave Creek Bridge, in Josephine County, is a good example of Oregon's covered bridges.

crib—A wooden box built to sit on the bottom of a waterway and hold gravel, rock, or concrete that will serve as the foundation for a bridge or other structure. If the crib is sunk into the ground, it is called a *caisson*.

crown—The highest point at the top of an *arch*.

deck—The surface part of a bridge that carries the *roadway* and sidewalks. Most bridge decks built in recent years are made of concrete. The deck typically is supported by a system of *stringers* and *floor beams*.

deck truss—A *truss bridge* with the traffic *roadway* located on top of its structure.

deflection—Change in shape of a structure due to bending, stretching, or shrinking under *load*.

diagonals—Sloping members of a *truss bridge* that connect between the top and bottom *chord*. The diagonals carry either *tension* or *compression*, but not both at the same time. They work with the *posts* and chords to carry the *loads* to the *piers*.

double-leaf bascule span—A *bascule bridge* with two leaves that open in the center and lift up to allow passage of river traffic. An example is McCullough's Old Young's Bay Bridge in Clatsop County.

drawbridge—Another word for a *movable bridge*.

ductile—Materials (rubber, some plastics, most steel) that stretch or *bend* before they break.

elastic/plastic—An elastic material changes shape and size when a *force* is applied, but returns to its original shape and size when the force is removed. Most materials used in bridges, such as steel, concrete and wood, remain elastic when the *stresses* applied are within normal range. For certain materials, when stresses exceed a certain level (called the yield strength or point), the material becomes plastic. This means the material does not return to its original shape and size and is permanently deformed.

engineer—There are four basic categories of engineers: civil, electrical, mechanical, and chemical. Engineers who design highway bridges must be licensed professional engineers, since they are responsible for making bridges safe for public use. Within the basic categories are many specialties. The field of civil engineering includes structural and geotechnical engineers. Structural engineers calculate the *loads* that a bridge must carry and work out the arrangement and size of the bridge *members* to carry the loads safely. Geotechnical engineers investigate ground conditions and advise about suitable *foundations*. *Movable bridges* require the expertise of mechanical and electrical engineers to design the machinery and electrical systems that operate the movable spans.

expansion joint—A meeting point designed to allow movement of bridge parts due to heat and cold.

falsework—Temporary structure used as support during construction.

fatigue—Fatigue is the gradual weakening of a material when a *force* is repeatedly applied and removed. A bridge must be able to withstand the forces on it without *bending* too much or breaking.

fill—Material (dirt, stone, boulders, concrete, sawdust) used to raise the ground level or to fill in space in structures.

fixed arch—An arch with no hinges.

flexibility—The opposite of stiffness, the measure of how much a structure will yield or flex under *load*.

floor beam—A bridge member running crosswise between the main *girders* or *trusses*. The floor beams support the *deck* system, particularly the *stringers*.

footing—A type of *foundation* for a bridge or other structure that sits on firm ground or rock and distributes *load* from the structure to the ground or rock.

force—A physical influence that moves or tries to move an object. Forces can be from the inside (internal) or from the outside (external). Force is exerted when one bridge member presses or pulls, pushes against, compresses, or twists another member. Engineers refer to external forces as *loads*. There are three internal forces in bridges: *tension, compression,* and *shear*.

gabion—A galvanized wire mesh box filled with stones and used to form an *abutment* or retaining wall.

galvanizing— To coat metal, especially iron or steel, with zinc to protect from decay (*rust*) by water or air. Named for Luigi Galvani (1737-98), an Italian who discovered certain chemicals can produce electronic action.

gate (traffic)—Barriers that swing up and down or in and out to prevent vehicles from proceeding onto the *lift span* of *movable bridges* during an opening.

girder—A girder is a large *beam* that spans between a structure's main supports. Bridge girders can be built of concrete, steel, or wood. The majority of Oregon's more than 6,500 highway bridges are girder bridges.

gusset plate—A metal plate used to connect multiple parts.

hanging deck truss—A *deck truss* bridge that hangs from supports at the top of the truss just below the level of the *roadway*.

hinged arch—An arch supported by a large *pin* at each end. (In a three-hinged arch, a third pin is located at the *crown* of the arch.)

iron/cast iron—As the main ingredient in *steel*, iron is especially useful in the construction of bridges. Cast iron, which is an alloy of iron and carbon, was used in the mid-nineteenth century in bridge construction, but was found to be too *brittle* to withstand large *tensile* forces.

lift span—The *movable* section of a *vertical lift* or *bascule bridge*.

load—A type of *force* that pushes or pulls on an object from the outside. Walking across a plank across a stream, you might feel the plank give or bounce because of your weight. This downward push is what engineers call load. Even when no one is standing on the plank (or on a bridge), it will sag because of the weight of the plank or bridge itself. The weight of the bridge is called the dead load and is permanent. Temporary loads on a bridge, such as people, vehicles, wind, and earthquakes are called live loads, as are rainwater, snow and ice.

main span—The longest *span* in a multi-span bridge located between the bridge's main *piers*, or *towers*.

member—One of the parts of a structure; the word is often used to refer to the parts of a truss bridge, for example, its *diagonals* and *chords*.

movable bridges—The three main movable bridge types are *swing, vertical lift*, and *bascule* and most fall into the *beam* or *truss* category.

multi-span bridges—Across wide valleys or over great rivers, a single-span bridge would be impossible or impractical. Multi-span bridges are carried by a series of *piers* or *bents* set in the river or valley. McCullough's major bridges on the Oregon Coast Highway are good examples of multi-span bridges.

parabola/parabolic—A curved shape, such as the curve of the main cables of a suspension bridge or the curve in many of McCullough's arch bridges.

pier—Bridge supports located in water.

piles/piling—Piling is a type of *foundation*, long, slender columns driven into the ground to support the structure's weight. It is made of various materials and shapes. In general, bridges built before World War II used timber piling, those after used *steel* or *precast concrete* piling. "Pile" generally refers to individual pieces; "piling" refers to a collection of pieces.

pin-connected—In nineteen- and early-twentieth-century *truss bridge* construction, truss members were joined by large metal pins. These bridges were referred to as "pin-connected" trusses. Later truss bridges used *rivets* or bolts.

plate-girder bridge—A bridge made with steel plate girders, large *beams* made from individual steel plates joined together with *rivets*, bolts or *welds*.

Portland cement—Concrete is made from Portland cement, water, sand, and gravel. Portland cement, named for the Isle of Portland in Dorset, England, is made out of burned lime and clay.

precast concrete—Concrete that has been cast into components before being placed in position. Generally, precast girders are *prestressed*.

prestressed concrete—Concrete put in a state of *compression* before *loads* are applied in order to minimize *tensile forces* in the concrete. Prestressing typically is done by stretching steel bars or wires, then transferring the stretching forces to the concrete.

reinforced concrete—A combination of concrete and reinforcing material. Concrete is very strong in *compression*, but not in *tension*. The most common reinforcing materials are steel bars, called reinforcing steel or rebar.

riprap—Large rocks placed to protect structures or embankments from moving water.

rivet—A metal pin with a head at each end that is passed through a hole to hold pieces of metal together. The head on one end of the rivet is formed once the rivet is through the hole.

rivet-connected bridges—These began replacing *pin-connected* bridges around the end of the nineteenth century. Riveting technology for bridges became obsolete around 1960, replaced by high strength bolts.

roadway—The part of the bridge that carries highway traffic, normally the top surface of the bridge's *deck*. Sidewalks on the bridge may be extensions of the roadway or separate structures.

saddle—Saddle-shaped receptacle for the main cable on top of a *suspension bridge tower*.

scour—Abrasive action of fast-moving water on the soil and rocks around bridge foundations that can undermine a footing and cause the bridge to fail.

shear—Combination of *forces* that causes parallel planes to slide relative to each other. In *beam* bridges, shear works together with *tension* and *compression* to support *loads* applied to the beam.

single-leaf bascule span—A *bascule bridge* with one leaf that lifts at one end to open and allow passage of traffic. McCullough's Lewis and Clark Bridge in Clatsop County is an example.

slab bridge—A type of *beam bridge* in which the *roadway deck* itself is designed to act as a wide beam. They are usually shorter than other bridge types, although they sometimes have spans as long as 80 feet. Slab bridges often are built by joining several *precast concrete* units together.

spall/spalling—Flaking or chipping of chunks of concrete off the surface of a structure, often caused by water working into fine cracks or holes in the concrete. The water can cause spalling by freezing and expanding or by rusting the reinforcing steel (also called rebar), which causes it to expand. In either case, expansion of the frozen water or rusted steel ruptures the concrete and pops it off the structure. Spall caused by salt air is particularly destructive.

span—The part of a bridge between two supports. There are single-span bridges and *multi-span bridges*. Span also refers to the distance between supports. The word may be used as a noun or verb.

spandrel arch bridge—An *arch bridge* with the *deck* located above the arch; the spandrel is the space between the arch and deck. In an open-spandrel arch bridge, the *roadway deck* is supported by the arch on a series of independent posts. In a closed-spandrel arch bridge, a solid wall exists between the arch and roadway deck.

starling—Pointed clusters of piling protruding from the upriver side of bridges, sometimes known as "dolphins." These additions act as fenders, preventing ships and river-borne debris such as logs and ice from crashing into and damaging bridge *piers*.

strain—When a bridge *member* has *forces* acting in it because of external *loads*, it changes length. The change in length is strain, which is expressed as a percentage of the original length.

strength—The load a structure can carry.

stress—Stress is *force* divided by area, such as pounds per square inch. Types of stresses are *tension, compression,* and *shear.*

stringer—One of a series of *beams* that directly support a bridge *deck.* Stringers usually run lengthwise with the bridge and are supported by *floor beams* running crosswise across the bridge.

substructure—The portion of a bridge that carries the weight of the *superstructure* to the ground. Substructures differ widely from bridge to bridge due to variations in ground conditions, bridge heights, superstructure designs, and budget.

superstructure—The horizontal portion of a bridge, which spans between the supports, or *substructure.* The superstructure carries the traffic and passes the weight of the traffic and superstructure to the substructure.

suspenders—Wire ropes or other components that support a bridge *deck* by suspending it from the main support structure overhead. On *suspension bridges,* the suspenders hang from the main cables that are draped between the *towers.*

suspension bridge—One of three main bridge types, along with *arch* and *beam* or *truss* bridges. On a suspension bridge, the weight of the *roadway deck* and traffic is supported by main cables draped over *towers.* The deck typically is connected to the main cables by *suspenders.* The main cables are tied to the ground by *anchorages.* Engineers have looked to this bridge type to span wide spaces since the mid-1800s.

swing bridge—One of three main movable bridge types, along with *vertical lift* and *bascule.* A swing bridge opens by rotating its span on a center *pier* located in the waterway. A good example is the central span of McCullough's Umpqua River Bridge at Reedsport, on the Oregon Coast Highway.

tension/tensile force—A force that tends to make something longer, or pull it apart.

through truss—A *truss bridge* with the traffic *roadway deck* located between the trusses.

thrust—A pushing *force* or pressure exerted by a structure or on a structure such as the downward and outward force exerted by an *arch* on each side.

tied arch—An *arch bridge* designed like an archer's bow, with the "string" (or tie) connecting one end to the other. This arrangement holds the *arch* in *compression* and the tie in *tension.* Tied arches are used at sites where it is not feasible to support the arch horizontally at ground level. The tie carries the horizontal *thrust* that would otherwise act on the *abutments.* McCullough used the tied arch, both in *reinforced concrete* and in steel, for several of his bridges.

torsion—A combination of *forces* that can be described as twisting. An example of torsion is the turning force required to twist a drill into a piece of wood. Torsion is produced in bridges by uneven gravity or wind *loads* on the *deck.*

tower—A tall *pier* or frame supporting the main cables of a *suspension bridge* or the inclined cables of a *cable-stayed bridge.*

truss bridge—Any braced framework may be called a truss. A truss is made up of *members* in a triangular arrangement. A truss bridge is a form of *beam bridge,* except that trusses are made up of smaller members. See also *deck truss, through truss.*

vertical lift bridge—One of three main movable bridge types, along with *swing* and *bascule.* A vertical lift bridge has a *span* that can be raised up and down to clear ships and boats. The lift span is raised and kept level by a system of synchronized machinery and *counterweights* similar to an elevator.

viaduct—A bridge over land, usually with multiple short or medium-length *spans*. McCullough's Cape Creek Bridge, on the Oregon Coast Highway, includes a two-tiered viaduct.

Warren truss—A type of *truss bridge* with *members* that form a series of triangles with alternating "A" and "V" shapes. The Warren truss dates to 1848.

weathering steel—A type of steel made with additives that produce a permanent coat of rust. The tight rust coat does not flake off, protecting the underlying steel from air and water. This means weathering steel does not need to be painted, and so is less expensive to maintain.

weld/welding—A method of heating the edges of separate pieces of metal until they melt and mix. When the edges cool, a solid connection forms called a joint. Part of the welding process involves adding metal to fill the gap between the two pieces.

✶ *Bibliography* ✶

Manuscript Collections and Oral Histories

Conde B. McCullough. Personal and Professional Papers. Held by Ray Allen, Portland, Oregon.

Conde B. McCullough. Personal Papers. Held by Mrs. John R. McCullough and John P. McCullough, Salem, Oregon.

Conde B. McCullough. Personal Photographs. Held by Mrs. John R. McCullough and John P. McCullough, Salem, Oregon.

John R. McCullough, son of Conde B. McCullough. Interview with Louis F. Pierce, 13 May 1980. Transcript from audio tape. Pierce's Collection, Junction City, Oregon.

Ivan Merchant, retired Oregon State Highway Department engineer. Interview by Louis F. Pierce, 4 June 1980. Transcript from audio tape. Pierce's Collection, Junction City, Oregon.

Ivan Merchant. Interview with Robert W. Hadlow and Richard L. Koochagian, 16 August 1990. Audio tape. Environmental Section, Technical Services, Oregon Department of Transportation.

P. M. Stephenson, retired Oregon State Highway Department engineer. Interview by Louis F. Pierce, 4 June 1980. Transcript from audio tape. Pierce's Collection, Junction City, Oregon.

Unpublished Papers, Theses, and Dissertations

Atly, Elizabeth Shellin. "C. B. McCullough and the Oregon Coast Highway Bridges Project." 1977. TMs.

"Engineering Antiquities Inventory." 1982. TMs. Environmental Section, Highway Division, Oregon Department of Transportation, Salem.

"Fact Sheet Eight: Rock Point Bridge." Southern Oregon Historical Society, Jacksonville, Oregon.

Hoyt, Jr., Hugh M. "The Good Roads Movement in Oregon: 1900-1920." Ph.D. diss., University of Oregon, 1966.

[Joachims, Larry] "Masonry Arch Bridges of Kansas." [1980]. TMs. Kansas State Historical Society, Topeka, Kansas.

Marston, Anson. "Good Roads." Address at Boone [County, Iowa] Chamber of Commerce. 31 March 1921. TMs. College of Engineering. Series 11/5/1. Iowa State University Archives, Ames.

———. "The Iowa State Highway Commission and Its Works." Address Delivered Before the Iowa League of Municipalities. 16 August 1921. TMs. College of Engineering. Series 11/5/1. Iowa State University Archives, Ames.

McGivern, James Gregory. "First Hundred Years of Engineering Education in the United States (1807-1907)." Ph.D. diss., Washington State University, 1960.

Mississippi Valley Conference of State Highway Departments. "Historical Highlights, 1909-1974." TMs. Copy in the Iowa Department of Transportation Library, Ames.

Rindell, John K. "From Ruts to Roads: The Politics of Highway Development in Washington State, 1899-1917." M.A. thesis, Washington State University, 1987.

Walker, H. B., and C. B. McCullough, "Effect of External Temperature Variation on Concrete Bridges." B.S. thesis, Iowa State College, 1910.

Archival Records

Ames, Iowa. Iowa State University Archives. Alumni Affairs Collection. Series 21/7/1.

Ames, Iowa. Iowa State University Archives. College of Engineering. Series 11/5/1.

Ames, Iowa. Iowa State University Archives. Commencement Exercises. "Program." 8 June 1916.

Ames, Iowa. Iowa State University Archives. Photograph Collection.

Ames, Iowa. State Board of Health. Certificates of Death for John Black McCullough, 25 October 1904. No. 94-0038; and Lenna Leota McCullough, 24 June 1912. No. 85-00366.

Ames, Iowa. State Census. Webster County, 1895.

Ames, Iowa. State Highway Commission. Standard Plans. Concrete Culverts. [by C. B. McCullough and T. H. MacDonald]. [1910s].

Des Moines, Iowa. Department of Public Health. Return of Marriage in the County of Dallas, for Conde B. McCullough and L. Marie Roddan, 4 June 1913.

Fort Dodge, Iowa. R. L. Polk and Co., City Directory, located at Fort Dodge Public Library. 1894-1906.

Corvallis, Oregon. Oregon State University Archives. *The Beaver*, Oregon State College Yearbooks, by Junior Classes of 1918, 1920.

Corvallis, Oregon. Oregon State University Archives. Honorary Degrees Files.

Corvallis, Oregon. Oregon State University Archives. Photograph Collection.

Portland, Oregon. State Health Division. Center for Health Statistics. Death Certificate of Conde B. McCullough, 6 May 1946. State File No. 3553.

Salem, Oregon. Department of State. Archives. Highway Division Records. Office Files. Accession No. 76A-90/3.

Salem, Oregon. Department of State. Archives. Highway Division Records. Contracts.

Salem, Oregon. Department of State. Archives. Highway Division Records. Correspondence of Chief Counsel. Accession 68A-34-1.

Salem, Oregon. Department of State. Archives. Highway Division Records. Newspaper Clipping Scrapbooks, 1916-1919, 1932-1950. Accession No. 78A-54.

Salem, Oregon. Department of Transportation. Office of General Files and History Center. Bridge Files.

Salem, Oregon. Department of Transportation. Technical Services. Bridge Section. Bridge Drawings.

Salem, Oregon. Department of Transportation. Technical Services. Bridge Section. Bridge Log of Oregon Highways.

Salem, Oregon. Department of Transportation. Technical Services. Bridge Section. Bridge Maintenance Files.

Salem, Oregon. Department of Transportation. Technical Services. Bridge Section. Historic Bridges Photograph Collection.

Salem, Oregon. Department of Transportation. Technical Services. Bridge Section. Job Record Cards.

Salem, Oregon. Department of Transportation. Technical Services. Bridge Section. Microfiche Contract Files.

Salem, Oregon. Department of Transportation. Technical Services. Environmental Section. Historic Bridge Files.

Salem, Oregon. Marion County. Probate Records. Case No. 12526. C. B. McCullough.

Salem, Oregon. St. Paul's Episcopal Church. Canonical Church Register, 1930s–1940s.

Salem, Oregon. Willamette University. *Bulletin.* 1926–29.

Suitland, Maryland. National Archives and Records Center. Washington National Records Center. Record Group 30, Records of the Bureau of Public Roads. Classified Central File.

Washington, D.C. Department of the Interior. National Park Service. Historic American Engineering Record. Oregon Historic Bridges Recording Project, 1990, deposited in the Library of Congress.

Articles

"Appeals Court Sustains Decision Against Luten Patents." *Engineering News-Record* 84 (26 February 1920): 417–18.

"Background of a State Highway Engineer, Howard W. Holmes of Montana participated actively in the advocacy of a state gasoline tax for Oregon—first state to adopt such a measure." *Pacific Builder and Engineer,* July 1941, 65.

Baldock, R. N. "Highway Design Applied to the State System." *Civil Engineering* 6 (October 1936): 634–46.

Bergendoff, R. N. "Notable Bridges in the United States." *Civil Engineering* 7 (May 1937): 321–25.

Billington, David P. "History and Esthetics in Concrete Arch Bridges." *Journal of the Structural Section, Proceedings of the American Society of Civil Engineers* 103, no. ST11 (November 1977): 2129–43.

Boehme, Christopher. "The Oregon Coast Bridges." *Pacific Northwest,* July 1988, 22.

"Bridge Construction Along the Inter-American Highway." *Bulletin of the Pan-American Union* 70 (1936): 63–64.

Burnham, John Chynoweth. "The Gasoline Tax and the Automobile Revolution." *Mississippi Valley Historical Review* 48 (December 1961: 435–59.

"Cantilever Erection Plan is Feature of Coos Bay Bridge." *Western Construction News* 11 (July 1936): 224–27.

Chase, O. C. "Design of Coast Highway Bridges." *Civil Engineering* 6 (October 1936): 647–51.

"Concrete Bridge with Longest Arch Completed in France." *Engineering News-Record* 92 (20 March 1924): 476-79.

"Current Construction Unit Prices, Oregon Coast Highway Bridges." *Engineering News-Record* 114 (20 June 1935): 897.

"Current News, U.S. to Build Superstructures of Pan-American Road Bridges." *Engineering News-Record* 115 (24 October 1935): 586.

"Decision of U.S. District Court of Iowa on Certain Luten Patents for Concrete Bridges." *Engineering and Contracting* 49 (23 January 1918): 94-96.

"A Discussion of the Administrative and Design Features of Highway Bridge and Culvert Work." *Engineering and Contracting* 42 (23 December 1914): 589-90.

"Editorial—Bridges and Earthquakes." *Engineering News-Record* 117 (31 December 1936).

"Editorial—A Novel Bridge Design." *Engineering News-Record* 88 (8 June 1922): 727.

"Editorial—Original Bridge Thought." *Engineering News-Record* 89 (2 November 1922): 727.

Eichinger, J. W. "Iowa's Largest State Job." *Iowa Engineer* 21 (October 1920): 1-3.

Eremin, A. A. "Freyssinet Arch Construction [in response to 'Designing First Freyssinet Arch to be Built in United States]." *Engineering News-Record* 108 (10 March 1932): 375-76.

Fahl, Ronald J. "S. C. Lancaster and the Columbia River Highway: Engineer as Conservationist." *Oregon Historical Quarterly* 74 (1973): 101-44.

Fletcher, Robert. "A Quarter Century of Progress in Engineering Education." In *Proceedings of the Fourth Annual Meeting, held in Buffalo, N.Y., August 20, 21, 22, 1896.* Society for the Promotion of Engineering Education. Lancaster, PA: New Era Printing Co., 1897.

"$400,000 Arterial at Portland." *Western Construction News and Highways Builder* 15 (June 1940): 200-201.

Freyssinet, Eugène. "Exposé d'Ensemble de l'Idée de Préconstrainte." *Annales de l'Institut Technique du Batiment et des Travaux Publics.* No. 13, Nouvelle série. Juin 1949.

————. "The Bridge at Villeneuve-sur-Lot, Department of Lot and Garonne, France: Advances in the Construction of Great Arches." Translated by Lloyd G. Frost. TMs. [1925]. Original, in French, appeared in *Le Génie Civil* 79 (1921).

————. "The 600-ft. Concrete Arch Bridge at Brest, France." Translated by S. C. Hollister. With discussion. *Proceedings of the American Concrete Institute* 25 (1929): 83-97, 98-99.

Gemeny, Albin L. "The Freyssinet Method of Concrete-Arch Construction." *Public Roads* 10 (October 1929): 148-150.

————. "Progress in Highway Design and Construction." *American Highways* 16 (October 1937): 23-27.

Gemeny, Albin L., and C. B. McCullough. "Freyssinet Method of Arch Construction Applied to Rogue River Bridge in Oregon." *Journal of the American Concrete Institute* 29 (October 1932): 57-79; letter by Charles S. White, 30 (February 1933): 301-04; authors' response, 30 (November-December 1933): 157-58.

Gemeny, Albin L., and C. B. McCullough. "Seven-Span Reinforced Concrete Arch Bridge." *Concrete and Construction Engineer* 28 (April 1933): 241-44.

Gottschalk, Otto. "Letter—Mechanical Methods of Stress Analysis." *Engineering News-Record* 106 (22 January 1931): 162.

Greene, W. K. "Special Problems of Hingeless Arch Erection [Henry Hudson Bridge in New York]." *Engineering News-Record*, 12 November 1936, 669-73.

Gronquist, C. H. "Simplified Theory of the Self-Anchored Suspension Bridge." With discussion by A. J. Meehan, C. B. McCullough, Jaroslav J. Polivka, William Bertwell, A. A. Eremin, and C. H. Gronquist. Paper No. 2151. *Transactions of the American Society of Civil Engineers* 107 (1942).

Growth [Groth], Karen. "Conde Balcom McCullough, an Engineer with Soul." *Historic Preservation League of Oregon Newsletter.* No. 41, Summer 1986. 5-7.

Hadlow, Robert W. "C. B. McCullough: The Engineer and Oregon's Bridge Building Boom, 1919-1936." *Pacific Northwest Quarterly* 82 (January 1991): 8-19.

————. "Oregon's Isaac Lee Patterson Memorial Bridge: The First Use of the Freyssinet Method of Concrete Arch Construction in the United States, 1932." *IA: The Journal of the Society for Industrial Archeology* 16, no. 2 (January 1990): 3-14.

Harris, Milton. "Types of Concrete Highway Spans." *Student Engineer [OAC]* 10 (1917): 23-26.

Hewes, L. I. "Highway Improvement Progress in the Western Mountain States." *American Highways* 17 (October 1938): 14-16.

Hill, Samuel. "Poor Roads Are the Costly Highways, not Good Ones." n.d. Clipping File. Bridge Section. Technical Services. Oregon Department of Transportation.

"Howard Holmes Streamlines Montana Highway Department." *Pacific Builder and Engineer,* July 1941, 47.

"Howard Holmes Succeeds Don McKinnon as Montana State Highway Engineer." *Pacific Builder and Engineer,* February 1941, 25.

"Internal Temperature Range in Concrete Arch Bridges. *Engineering and Contracting* 40 (12 November 1913): 533.

[Joachims, Larry] "Marsh Arch Bridges: A Part of Kansas Transportation History." *Kansas Preservation* [newsletter, Historic Preservation Department, Kansas State Historical Society] 5 (March-April 1983): 1-5.

Kelley, E. F. "Steel Bridge Standards of the Iowa Highway Commission." *Engineering Record* 70 (12 December 1914): 631-32.

Learned, Edmund P. "Gasoline Taxes: Theory, Practice, and Hazards." *Engineering News-Record* 104 (2 January 1930): 12-16.

"Lower Court Rules Against Luten Patent." *Engineering News Record* 80 (17 January 1918): 144.

"Luten Patent Decision Upheld in Higher Court." *Engineering News-Record* 81 (26 December 1918): 1200.

"Luten Patents on Concrete Construction." *Engineering News* 71 (5 February 1914): 329.

MacDonald, Thomas H. "Bridge Patent Litigation in Iowa." *Iowa Engineer* 18 (January 1918): 118ff.

————. "Contrasting United States and European Practices in Road Development." *American Highways* 18 (January 1939): 10-14.

————. "Highway Administration in the United States." *Good Roads* 68 (November 1925): 277.

————. "The Highway Recovery Program." *American Highways* 12 (October 1933): 3-5, 27.

"Marsh Arch Bridges as Part of Kansas' Transportation History." *Kansas Preservation* 5 (March-April 1983): 1-5.

[Marston, Anson] "A Discussion of the Administrative and Design Features of Highway and Culvert Work." *Engineering and Contracting* 22 (23 December 1914).

Marston, Anson. "A Word To Freshman Engineers." *Iowa Engineer* 21 (October 1920): 4.

———. "The Choice of Subjects for Theses." *Iowa Engineer* 11 (April 1911): 315–16.

———. "The Civil Engineer and His Place in the World's Economy." *Iowa Engineer* 9 (November 1909): 196–202.

———. "The Engineer as Citizen." reprint. *Bulletin of the Associated State Engineering Societies*, October 1929.

———. "The Engineering Division." *Iowa State Alumnus* 19 (June 1924): 276–78.

———. "Freshmen Engineers." *Iowa Engineer* 15 (October 1914): 5.

———. "Highway Engineering in Iowa." *Iowa Engineer* 14 (October 1913): 15–19.

———. "History and Organization of Engineering Experiment Station." In Iowa Engineering Experiment Station. *Bulletin* 8 (November 1908): 252–56.

———. "Legislative Appropriation for Engineering Division." *Iowa Engineer* 9 (July & September 1909): 150–51.

———. "The State's Responsibility in Road Improvement." *Iowa Engineer* 7 (November 1908): 208–15.

———. "What is Engineering." *Iowa Engineer* 21 (May 1921): n.p.

McCullough, C. B. "An Analysis of the Highway Tax Structure In Oregon." *American Highways* 17 (October 1938): 3–7.

———. "Arch Bridge Ribs Anchored with Concrete Keys." *Engineering News-Record* 83 (27 November–4 December 1919): 924.

———. "Are the Highway Commission Bridges Too Heavy?: County vs. State Control." In Iowa Highway Commission *Service Bulletin No. 12*, December 1914, 3–6.

———. "Bridging the Rio Chiriqui on the Pan-American Highway." *Engineering News-Record* 117 (26 November 1936): 757–59.

———. "Concrete Highway Bridge Construction as Standardized by Iowa Commission." *Engineering Record* 70 (7 November 1914): 514–17.

———. "Concrete Under the Microscope." *Student Engineer [OAC]* 10 (1917): 16–22.

———. "Cost Economics in Concrete Bridges," in R. W. Crum, ed., *Proceedings of Annual Meeting of Highway Research Board*. vol. 10. Washington, DC: National Research Council, 1931.

———. "Derivation of Theories Underlying Mechanical Methods of Stress Analysis." *Engineering News-Record* 105 (25 September 1930): 489–90.

———. "Design Graphs for Highway Suspension Bridges of Moderate Span," in *Proceedings of the Annual Meeting of the Highway Research Board*. vol. 20. Washington, DC: National Research Council, Division of Engineering and Industrial Research, 1940.

———. "Design of Concrete Bowstring-Arch Bridge, Including Analysis of Theory." *Engineering News-Record* 107 (27 August 1931): 337–98.

[McCullough, C. B.] "The Design of Concrete Highway Bridges with Special Reference to Standardization." *Engineering and Contracting* 43 (24 March 1915): 268–70.

McCullough, C. B. "Designing Highways to Meet Pedestrian Needs." *Roads and Bridges* 83 (November 1945): 60–61, 94, 96, and 98.

———. Discussion of Donald N. Becker, "Development of the Chicago Type Bascule Bridge." *Transactions of the American Society of Civil Engineers* 109 (1944): 1035-43 (article, 995-1025).

———. Discussion of "Fundamental Aspects of the Depreciation Problem—A Symposium," in *Transactions of the American Society of Civil Engineers* 108 (1943): 1289-93 (article, 1236-61).

———. Discussion of C. H. Gronquist, "Simplified Theory of the Self-Anchored Suspension Bridge," in *Transactions of the American Society of Civil Engineers* 107 (1942): 955-75.

———. Discussion of H. E. Langley and Edward J. Ducey, "Reconstruction of the Walpole–Bellows Falls Arch Bridge," in *Transactions of the American Society of Civil Engineers* 105 (1940): 1702-7 (article, 1675-1700).

———. Discussion of "A Symposium on the Economics of Low Cost Highway Bridges," in *Proceedings of Annual Meeting of Highway Research Board.* vol. 11. pt. 1. Washington, DC: National Research Council, 1931 (pp. 114-23).

———. Discussion of Wilbur M. Wilson, "Laboratory Tests of Multiple-Span Reinforced Concrete Arch Bridges," in *Transactions of the American Society of Civil Engineers* 100 (1935): 424-54.

———. "Economics of Highway Alignment Design." *Proceedings of the Highway Research Board.* vol. 21. Washington, DC: National Research Council, 1941, pp. 164-76.

———. "Evaluating Highway Extensions." *Engineering News-Record* 120 (3 March 1938): 530-33, and discussion in 120 (14 April 1938): 528.

———. "Flexural Resistance of Shallow Concrete Beams." *Engineering News-Record* 115 (19 September 1935): 406-7, see also discussion on 115 (31 October 1935): 617; 18 (7 November 1935): 648; and 115 (21 November 1935): 722-23.

———. "Foundations for Highway Bridges." *Roads and Streets* 67 (March 1927): 313-15.

———. "Fourth Street Viaduct." *Engineering News-Record* 113 (4 October 1934): 429.

———. "Gaining First Cost Economics in Concrete Highway Bridges." *Concrete* 38 (April 1932): 43-44.

———. "Grade Line Treatment for Highway Bridges." *Roads and Streets* 67 (November 1927): 494-96.

———. "Gunite Retains Integrity on Oregon Road Bridges." *Engineering News-Record* 111 (31 August 1933): 259-60.

———. "Help for the Pedestrian in Rural Roadway Design." *Better Roads* 15 (July 1945): 18-20.

———. "Highway Costs and Earnings." *American Highways* 19 (July 1940): 8-14.

———. "How Oregon Builds Highway Bridges." *Oregon Motorist* 10 (February 1930): 13-15, 27.

———. "Highway Viaducts in Oregon of Prefabricated Timber. *Engineering News-Record* 113 (4 October 1934): 429-30.

———. "Investigation of Foundations in Highway Bridge Surveys." *Roads and Streets* 68 (August 1928): 385-91.

———. "Large Steel Arch Ribs Encased in Gunite." *Engineering News-Record* 88 (8 June 1922): 942-45.

———. "Long Range Plan for Salem Oregon." *Pacific Builder and Engineer* 51 (September 1945): 45-47.

————. "Maintaining Oregon's Highway Bridges: The Methods Used by the Oregon State Highway Department Described in a Paper Presented at the Recent Meeting of the American Association of State Highway Officials." *Roads and Streets* 67 (March 1927): 115-20.

————. "Maintenance and Repair of Bridges." *Journal of the American Concrete Institute* 35 (February and June 1939): 229-56.

————. "Modern Design and Construction Practice for Wide-Span Arches in U.S.A." Assn Internationale des Ponts et Charpertes, *Memoires* 6 (1940-41): 189-209.

————. "New Design Principle in Creosoted Timber Bridges," in *Proceedings of the 32nd Annual Meeting of the American Wood Preservers' Association.* 1936.

————. "Old Suspension Bridge Used in Erecting New Arch." *Engineering News-Record* 89 (2 November 1922): 730-33.

————. "Oregon Plans Expansion Joint Study." *Engineering News-Record* 125 (1 August 1940): 168-69.

————. "Oregon Steel Arch Bridge Erected by Cableway." *Engineering News-Record* 96 (13 May 1926): 760-62.

————. "Oregon Tests on Composite (Timber-Concrete) Beams." *Journal of the American Concrete Institute* 14 (April 1943): 429-40.

————. "Oregon's Merit System for Highway Personnel Administration." *Western Construction News and Highways Builder* 15 (July 1940): 236-39.

————. "Remarkable Series of Bridges on Oregon Coast Highway." *Engineering News-Record* 115 (14 November 1935): 677-79.

————. "Rogue River Bridge at Gold Beach, Oregon: Series of Reinforced Concrete Arches Using Freyssinet Method of Decentering and Arch Adjustment." *Western Construction News* 6 (10 July 1931): 341-42.

————. "Self-Liquidating Plan for Oregon's Coast Highway Bridges." *Engineering News-Record* 114 (6 June 1935): 814-15.

————. "16th Century Engineering in Central America." *Western Construction News and Highways Builder* 11 (August 1936): 259-60.

————. "Standard of Design for Strategic Roads." *Pacific Builder and Engineer* 47 (October 1941): 48, 52-53, and 74.

————. "Tax Structure Analysis for Highway Planning." *Civil Engineering* 8 (August 1938): 534-36.

————. "Timber Bridges in Oregon," in *Proceedings of the Highway Research Board* 23 (1943): 235-50.

————. "Timber Highway Bridges in Oregon." *Engineering News-Record* 109 (25 August 1932): 213-14.

————. "Truss Centering Used for 113-Ft. Concrete Arch." *Engineering News-Record* 84 (29 April 1920): 851-52.

————. "Walkers are Difficult to See." *The American City* 61 (January 1946): 111-13.

————. "Western Practice Utilizes New Types" (part of symposium on "Modern Short-Span Bridges"). *Civil Engineering* 2 (September 1932): 549-62.

————. "Where Cost Economies Are Gained in Concrete Highway Bridges." *Concrete* 38 (March 1931): 45-46.

————, and Raymond Archibald. "Bridging the Rio Choluteca with a Two-Span Suspension Structure." *Engineering News-Record* 118 (21 January 1937): 87-88.

————, and Raymond Archibald. "Self-Anchored Eyebar Cable Bridge." *Engineering News-Record* 118 (17 June 1937): 912-13.

————, and John Beakey. "The Economics of Highway Planning Propounded by Oregon Engineers." *Roads and Streets* 80 (October 1937): 61-72.

————, and Albin L. Gemeny. "Designing the First Freyssinet Arch to Be Built in the United States." *Engineering News-Record* 107 (26 November 1931): 841-45.

————, and H. G. Smith. "Oregon Uses Lambert Conformal Conic Projection in Highway Surveys." *Civil Engineering* 15 (May 1945): 209-12.

McGaffey, Ernest. "What Federal Aid Means to the West." *Sunset Magazine* 56 (March 1926): 36-37, 94-95.

"Mr. Luten's Attorney Takes Exception." *Engineering News-Record* 84 (17 June 1920): 1220-21.

"Montana Adds Two Bridges on Important Highway Routes." *Western Construction News and Highways Builder* 11 (August 1936): 246-48.

Murrow, Lacey V. "Letter—Obsolescence in Highway Design [favorable response to Baldock article]." *Civil Engineering* 6 (November 1936): 764.

National Cyclopædia of American Biography. vol. 4. S.v. "Fuertes, Estévan Antonio."

National Cyclopædia of American Biography. vol. 35. S.v. "Marsh, James B."

National Cyclopædia of American Biography. vol. 44. S.v. "Marston, Anson." [1864-1949]

National Cyclopædia of American Biography. vol. 35. S.v. "McCullough, Conde B." [1887-1946]

Nichols, C. S., and C. B. McCullough. "Results of Experiments to Determine the Internal Temperature Range in Concrete Arches" (Abstract of Bulletin No. 30, Engineering Experiment Station, Iowa State College of Agriculture and Mechanic Arts). *Engineering and Contracting* 40 (12 November 1913): 533.

"A Novel Bridge Design." Editorial. *Engineering News-Record* 88 (8 June 1922): 940.

"Oregon Highway Bridge to Be Built by Freyssinet Arch Method." *Engineering News-Record* 105 (21 August 1930): 290.

"Original Bridge Thought." Editorial. *Engineering News-Record* 89 (2 November 1922): 727.

"The Organization and Standards of the Iowa Highway Commission." *Engineering and Contracting*, 15 July 1914.

Page, Logan Waller. "Good Roads and How to Build Them." *Scientific American* 106 (16 March 1912): 236-38.

Paxson, Glenn S. "Construction of Coast Highway Bridges." *Civil Engineering* 6 (October 1936): 651-55.

————. "Maintenance and Repair of Concrete Bridges on the Oregon Highway System. *Journal of the American Concrete Institute*, vol. 42 of *Proceedings*, 17 (November 1945): 105-14.

————. "Memoir of Conde Balcom McCullough." *Transactions of the American Society of Civil Engineers* 112 (1947): 1489-91.

————, and Marshall Dresser. "Concrete Arch Ribs of Rogue River Bridge Decentered by Built-In Jacks." *Construction Methods* 15 (April 1933): 36-39.

Peabody, L. E. "The Western States Traffic Survey." *Public Roads: A Journal of Highway Research* 13 (March 1932): 1-18.

Pierce, Louis F. "Esthetics in Oregon Bridges—McCullough to Date." *ASCE Preprint* American Society of Civil Engineers Convention, Portland, Oregon, 14-18 April 1980, no. 80-026.

Purcell, Charles H. "Sketching Construction History of San Francisco-Oakland Bay Bridge." *Western Construction News and Highways Builder* 11 (November 1936): 353-54.

"Rainbow Arch Adds Variety to Kansas Highways." *Kansas Preservation* 2 (November-December 1980): 1-2.

Reed, E. T. "O.S.C. President for 25 Years." *Oregon State Monthly* 11 (June 1931): 3-8.

Reed, M. E. "Building Yaquina Bay Bridge on Oregon Coast Highway." *Western Construction News and Highways Builder* 11 (May 1936): 133-36.

Richards, Carl Price. "Design and Construction of the Bridges." *Souvenir Program: Oregon City—West Linn Bridge.* n.p., 1982.

Ricketts, E. G. "Safeguarding Bridges by Inspection." *Engineering News-Record* 121 (14 July 1938): 51-54.

"Rogue River Bridge Registered as National Landmark." *The Oregon Civil Engineer* 30 (January 1983): 1.

Sanders, S. J. "Construction Review, Bridges and River and Harbor Work". *Western Construction News,* 10 March 1932, 143-45.

Scott, W. A. "Rogue River Bridge at Gold Beach, Oregon." *Western Construction News and Highways Builder* 7 (25 May 1932): 281-82.

Seely, Bruce E. "The Scientific Mystique in Engineering: Highway Research at the Bureau of Public Roads, 1918-1940." *Technology and Culture* 25 (October 1984): 798-831.

Slack, Searcy B. "Measuring Strain and Temperature in a 160-Ft. Concrete Arch Bridge." *Engineering News-Record* 29 August 1929, 336-39.

Soth, Lauren K. "He Pulled Iowa Out of the Mud." Iowa State College *Alumnus,* October 1931, 1-3.

Staff Article. "The Organization and Standards of the Iowa Highway Commission. *Engineering and Contracting* 42 (15 July 1914): 55-63.

Staff Article. "Standard I-Beam and Pile Highway Bridges of the Iowa State Highway Commission." *Engineering and Contracting* 42 (29 July 1914): 102-4.

"Standard I-Beam and Pile Highway Bridges of the Iowa State Highway Commission." *Engineering and Contracting,* 29 July 1914.

"State Control of Highway Bridge Construction." Editorial. *Engineering News* 67 (13 June 1912): 1137-38.

"Steel Bridge Awarded Prize Is of Three-Hinged Tied-Arch Type." *Engineering News-Record* 113 (23 August 1934): 248.

Thompson, J. T. "Freyssinet Method of Arch Construction." *Baltimore Engineer* 5 (January 1931): 4-6.

———. "Stresses Under the Freyssinet Method of Concrete-Arch Construction. *Engineering News-Record* 105 (21 August 1930): 291.

Turner, C. A. P. "Bridge Economics." *Engineering News-Record* 113 (20 December 1934): 801.

"Two Interesting Concrete Bridges in Oregon." *Engineering and Contracting* 56 (26 October 1921): 389-91 [bridges over Hood River and Rogue River].

"Unique Construction Procedure on Long-Span Concrete Bridge Arches at Brest, France." *Engineering News-Record* 103 (31 October 1929: 691-95.

Watson, Wilbur. "Bridges and Civilization." *American Highways* 12 (October 1933): 6-11.

White, F. R., and J. H. Ames. "Is a County Engineer Necessary." *Iowa Engineer* 13 (February 1913): 279-90.

Whitney, Charles S. "Long Span Concrete Arch Design in France." *Engineering News-Record* 93 (18 September 1924): 463-65.

"Who's Who Among Ames Men," *Iowa Engineer* 26 (December 1925): 14.

Who Was Who in America, 1951-60. S.v. "MacDonald, Thomas Harris."

Who Was Who in America, 1951-60. S.v. "Purcell, Charles Henry."

Young, C. R. Review of *Elastic Arch Bridges*, by Conde B. McCullough and Edward S. Thayer. In *Canadian Engineer*, 31 May 1932, n.p.

Newspapers

Astoria (Oregon) *Astorian-Budget*. 1932-47, Clipping File Scrapbooks, Oregon State Archives, Salem, Accession No. 78A-54 [CF-OSA].

Corvallis (Oregon) *Gazette-Times*. 1932-47, CF-OSA.

Cottage Grove (Oregon) *Sentinel*. 1933, CF-OSA.

Eugene (Oregon) *News*. 1932-47, CF-OSA.

Eugene (Oregon) *Register-Guard*. 1932-47, CF-OSA.

Forest Grove (Oregon) *Washington County News-Times*. 1946, CF-OSA.

Fort Dodge (Iowa) *Messenger*. 1904, 1912, 1913.

Gold Beach (Oregon) *Curry County Reporter*. 1932-50, CF-OSA; 1982.

Marshfield/Coos Bay (Oregon) *Coos Bay Times*. 1932-47, CF-OSA.

Newport (Oregon) *Journal*. 1932-47, CF-OSA.

North Bend (Oregon) *Harbor*. 1932-37, CF-OSA.

Oregon Agricultural College *Barometer*. 1914-46.

Oregon City Enterprise. 1934, CF-OSA.

Oregon Voter. 5 May 1934.

Portland *Journal of Commerce*. 1933-46, CF-OSA.

Portland *Oregon Journal*. 1913-31; 1932-46, CF-OSA.

Portland *Oregonian*. 1913-31; 1932-46, CF-OSA.

Reedsport (Oregon) *Courier*. 1932-37, CF-OSA.

Roseburg (Oregon) *Review*. 1887, 1889.

Salem *Capital Journal*. 1932-51, CF-OSA.

Salem *Capital Press*. 1932-37, CF-OSA.

Salem *Oregon Statesman*. 1861; 1913-31; 1932-50, CF-OSA.

Springfield (Oregon) *News*. 1932-47, CF-OSA.

Washington D.C. *U.S. Daily*. 31 May 1932.

Periodicals

Construction Methods. 1910-40.

Oregon Motorist. vols. 1-20 (1920-40), especially May 1936.

Proceedings. American Road Builders' Association. 1903-40.

Proceedings. American Society of Civil Engineers. 1910-40.

Proceedings of the Annual Meeting of the Highway Research Board. Highway Research Board. National Research Council. National Academy of Sciences. Washington, 1920-45.

Public Roads: A Journal of Highway Research. Bureau of Public Roads. U.S. Department of Agriculture. vols. 1-21 (1919-40).

Transactions. American Society of Civil Engineers. 1910-1947.
Western Construction News and Highways Builder. 1915-40 [under a variety of names].

Books

Agg, Thomas R., and John E. Brindley. *Highway Administration and Finance.* New York: McGraw-Hill Book Co., 1927.

Agg, Thomas Radford, and C. B. McCullough. *An Investigation of Concrete Roadways.* Ames, Iowa: Tribune Publishing Company, 1916.

American Association of State Highway Officials. *AASHO: The First Fifty Years, 1914-1964.* Washington, DC: 1965.

Andrew, Charles E. *Final Report on Tacoma Narrows Bridge, Tacoma, Washington.* [Olympia]: [Washington Toll Bridge Authority], 1952.

Baldock, R. H., and C. B. McCullough. *The Merit System for Engineering Personnel.* Technical Bulletin No. 9. Salem: Oregon State Highway Commission, 1938.

Beckham, Stephen Dow. *Land of the Umpqua: A History of Douglas County, Oregon.* Roseburg, OR: Douglas County Commissioners, 1986.

Bill, Max. *Robert Maillart: Bridges and Constructions.* 3rd Edition. New York: Frederick A. Praeger, Publishers, 1969.

Billington, David P. *Robert Maillart's Bridges: The Art of Engineering.* Princeton, NJ: Princeton University Press, 1979.

Bishop, Morris. *A History of Cornell.* Ithaca, NY: Cornell University Press, 1962.

Blair, Karen. *The History of American Women's Voluntary Organizations, 1810-1960.* Boston: G. K. Hall and Co., 1989.

Bleich, Friedrich, Conde B. McCullough, Richard Rosencrans, and George S. Vincent. *The Mathematical Theory of Vibration in Suspension Bridges.* Bureau of Public Roads, Department of Commerce. Washington, DC: Government Printing Office, 1950.

Brindley, John E. *History of Road Legislation in Iowa.* Iowa Economic History Series. Iowa City: The State Historical Society of Iowa, 1912.

Buenker, John D., John C. Burnham, and Robert M. Crunden. *Progressivism.* Cambridge, MA: Schenkman Publishing Co., 1977.

Condit, Carl W. *American Building: Materials and Techniques from the Beginning of the Colonial Settlements to the Present.* 2nd ed. Chicago: University of Chicago Press, 1982.

Directory of Engineering Alumni of Iowa State College. Ames: The Iowa Engineer, 1938.

Douthit, Nathan. *A Guide to Oregon South Coast History.* Coos Bay, OR: River West Books, 1986. New edition Corvallis, OR: Oregon State University Press, 1999.

Franck, Harry A., and Herbert C. Lanks. *The Pan American Highway: From the Rio Grande to the Canal Zone.* New York: D. Appleton-Century Co., 1940.

Frankland, F. H. *Suspension Bridges of Short Span.* [New York]: American Institute of Steel Construction, 1934.

Fuller, Almon. *A History of Civil Engineering at Iowa State College.* Ames: Alumni Achievement Fund of Iowa State College, 1959.

Fuller, Wayne. *RFD: The Changing Face of Rural America.* Bloomington: Indiana University Press, 1964.

Gemeny, Albin L., and C. B. McCullough, *Application of Freyssinet Method of Concrete Arch Construction to the Rogue River Bridge in Oregon: A Cooperative Research Project by the U.S. Bureau of Public Roads and Oregon State Highway Commission.* Salem: Oregon State Highway Commission, 1933.

Giedion, Siegfried. *Space, Time and Architecture.* Cambridge, MA: Harvard University Press, 1967.

Gilkey, Herbert J. *Anson Marston: Iowa State University's First Dean of Engineering.* Ames: Iowa State University, College of Engineering, 1968.

Gray, L. *Souvenir of Yaquina Bay Bridge Dedication and the Completion of the Last Link in the Oregon Coast Highway.* Salem, OR: Capital City Binding, [1936].

Haber, Samuel. *Efficiency and Uplift, Scientific Management in the Progressive Era, 1890-1920.* Chicago: University of Chicago Press, 1964.

Harding, T. Swann. *Two Blades of Grass: A History of Scientific Development in the U.S. Department of Agriculture.* Norman, OK: 1947.

Hays, Samuel P. *Conservation and Gospel of Efficiency: The Progressive Conservation Movement, 1890-1920.* Cambridge, MA: Harvard University Press, 1959.

Hewett, Waterman Thomas. *Cornell University: A History.* Volume 2. New York: The University Publishing Society, 1905.

Highway Research Board. *Ideas and Actions: A History of the Highway Research Board, 1920-1970.* Washington, DC: National Research Council, National Academy of Sciences, [1971].

History of the Reformed Presbyterian Church of America. n.p., n.d., located at Presbyterian Historical Society, Philadelphia, PA.

Hool, George A., and W. S. Kinne. *Movable and Long-Span Steel Bridges.* New York: McGraw-Hill Book Co., 1923.

Hopkins, H. P. *A Span of Bridges: An Illustrated History.* New York: Praeger, 1970.

Hopkins, Harry L. *Spending to Save: The Complete Story of Relief.* New York: W. W. Norton & Company Inc., 1936.

Jackson, Donald C. *Great American Bridges and Dams,* with foreword by David McCullough, Great American Places Series. Washington: The Preservation Press, 1988.

Ketchum, Milo Smith. *The Design of Highway Bridges and the Calculation of Stress in Bridge Trusses.* New York: McGraw-Hill Book Co., 1908.

Kirkham, John Edward. *Highway Bridges: Design and Cost.* New York: McGraw-Hill Book Co., 1932.

———. *Reinforced Concrete: Theory and Design.* New York: McGraw-Hill Book Co., 1941.

———. *Structural Engineering.* 2nd ed. New York: McGraw Hill Book Co., 1933.

Layton, Edwin T., Jr. *The Revolt of the Engineers: Social Responsibility and the American Engineering Profession.* Cleveland: The Press of Case Western Reserve University, 1971.

McCullough, C. B. *Design of Waterway Areas for Bridges and Culverts.* Technical Bulletin No. 4. Salem: Oregon State Highway Commission, 1934.

———. *Determination of Highway System Solvencies.* Technical Bulletin No. 8. Salem: Oregon State Highway Commission, 1937.

———. *Economics of Highway Bridge Types.* Chicago: Gillette Publishing, 1929.

———. *Highway Bridge Location.* Department of Agriculture Department Bulletin No. 1486. Washington, DC: Government Printing Office, 1927.

————. *Highway Bridge Surveys.* Department of Agriculture Technical Bulletin No. 55. Washington, DC: Government Printing Office, 1928.

————. *Izbor Polozaja Drumskih Mostova.* [Location Choices for Country Bridges]. Kraljevina Jugoslavija. Ministarstvo gradevina. Uz saradnju Biroa za drzvae putove Ministarstva poljoprivrede Sjedinjenih Drzava. Prevod sa engleskog. Beograd, Stampano u grafickoj radionici Min. gradevina, 1930. Translation of McCullough, C. B. *Highway Bridge Location.* Department of Agriculture Department Bulletin No. 1486. Washington, DC: GPO, 1927.

————, and John Beakey. *The Economics of Highway Planning.* Technical Bulletin No. 7. Salem: Oregon State Highway Commission, 1937.

————, ————, and Paul Van Scoy. *An Analysis of the Highway Tax Structure in Oregon.* Technical Bulletin No. 10. Salem: Oregon State Highway Commission, 1938.

————, Albin L. Gemeny, and W. R. Wickerham. *Electrical Equipment on Movable Bridges.* Department of Agriculture Technical Bulletin No. 265. Washington, DC: Government Printing Office, 1931.

————, and John R. McCullough. *The Engineer at Law: A Resumé of Modern Engineering Jurisprudence.* With forewords by Hon. James T. Brand and Hon. J. M. Devers. 2 vols. Ames: The Iowa State College Press, issued by the Collegiate Press, Inc., in cooperation with the Oregon State Highway Department, 1946.

————, and Glenn S. Paxson. *Effect of Heavy Motor Transport on Highway Bridge Stresses.* Technical Bulletin No. 6. Salem: Oregon State Highway Commission, 1937.

————, ————, and Richard Rosecrans. *The Experimental Verification of Theory for Suspension Bridge Analysis (Fourier-series Methods).* Technical Bulletin No. 15. Salem: Oregon State Highway Commission, 1942.

————, ————, and ————. *Multiple-span Suspension Bridges; Development and Experimental Verification of Theory.* Technical Bulletin No. 18. Salem: Oregon State Highway Commission, 1944.

————, ————, and Dexter R. Smith. *Derivation of Design Constraints for Suspension Bridge Analysis.* Technical Bulletin No. 14. Salem: Oregon State Highway Commission, 1940.

————, ————, and ————. *An Economic Analysis of Short-span Suspension Bridges for Modern Highway Loadings.* Technical Bulletin No. 11. Salem: Oregon State Highway Commission, 1938.

————, ————, and ————. *Rational Design Methods for Short-span Suspension Bridges for Modern Highway Loadings.* Technical Bulletin No. 13. Salem: Oregon State Highway Commission, 1940.

————, and Edward S. Thayer. *Elastic Arch Bridges.* New York: John Wiley and Sons, 1931.

Merriam, Lawrence C., and David G. Talbot. *Oregon's Highway Park System 1921-1989: An Administrative History.* Salem: Oregon Parks and Recreation Department, 1992.

Mock, Elizabeth. *The Architecture of Bridges.* Reprinted edition. New York: Museum of Modern Art, 1949, 1972.

National Academy of Sciences. National Research Council. Highway Research Board. *Ideas and Action: A History of the Highway Research Board, 1920-1970.* Washington, DC: n.p., [1971].

Nichols, C. S. *Iowa State College of Agriculture and Mechanic Arts Directory of Graduates of the Division of Engineering.* Ames: n.p., 1912.

————, and Conde B. McCullough. *The Determination of Internal Temperature Range in Concrete Arch Bridges*. Engineering Experiment Station Bulletin No. 30. Ames: Iowa State College of Agriculture and Mechanic Arts, 1913.

Petroski, Henry. *Engineers of Dreams: Great Builders and the Spanning of America*. New York: A. A. Knopf, 1995.

Plowden, David. *Bridges: The Spans of North America*. New York: W. W. Norton & Company, 1974.

Rezneck, Samuel. *Education for a Technological Society: A Sesquicentennial History of the Rensselaer Polytechnic Institute*. Troy, NY: Rensselaer Polytechnic Institute, 1968.

Rose, Mark H. *Interstate: Expressway Politics, 1939-1989*. revised edition. Knoxville: University of Tennessee Press, 1990.

Ross, Earle D. *Democracy's College: The Land-Grant Movement in the Formative Years*. Ames: The Iowa State College Press, 1942.

————. *A History of the Iowa State College of Agriculture and Mechanic Arts*. Ames: The Iowa State University Press, 1942.

Sealy, Antony. *Bridges and Aqueducts*. London: Hugh Evelyn Limited, 1976.

Seely, Bruce E. *Building the American Highway System: Engineers as Policy Makers*. Philadelphia: Temple University Press, 1987.

Smith, Dwight A. *Columbia River Highway Historic District: Nomination of the Old Columbia River Highway in the Columbia Gorge to the National Register of Historic Places*. Salem: Environmental Section, Technical Services Branch, Oregon State Highway Division, Oregon Department of Transportation, 1984.

Stewart, George R. *N.A. 1—The North-South Continental Highway—Looking South*. Boston: Houghton Mifflin Co., and Cambridge, MA: The Riverside Press, 1957.

Taylor, Frederick Winslow. *The Principles of Scientific Management*. New York: Harper and Brothers Publishers.

Timoshenko, Stephen P. *History of Strength of Materials*. New York: McGraw-Hill Book Co., 1953, reprint ed. New York: Dover Publications, 1983.

Van der Zee, John. *The True Story of the Design and Construction of the Golden Gate Bridge*. New York: Simon and Schuster, 1986.

Washington Department of Highways and the Oregon Department of Transportation, *Report on Trans-Columbia River Interstate Bridge Studies*. Technical Bulletin No. 16. Salem: Oregon State Highway Commission, 1944.

Watson, Ralph. Compiler. *Casual and Factual Glimpses at the Beginning and Development of Oregon's Roads and Highways*. Salem: Oregon State Highway Commission, 195[?].

Wiebe, Robert H. *The Search for Order, 1877-1920*. New York: Hill and Wang, 1967.

Williams, J. Kerwin. *Grants-in-Aid Under the Public Works Administration: A Study in Federal-State-Local Relations*. New York: Columbia University Press, 1939.

Government Documents

California. Division of Highways. *Report*. 1917-1945.

Iowa. State Highway Commission. *Annual Report*. 1913-1940.

Iowa. State Highway Commission. *Service Bulletin*. August 1916 issue.

Iowa. Iowa State College. *Bulletin*. 1892-1920.

Iowa. Iowa State College. *General Catalogue.* 1916.

Michigan. State Highway Commission. *Biennial Report.* 1901–1942.

Montana. State Highway Department. *History of the Montana State Highway Department, 1913-1942.* Helena: 1943.

North Carolina. State Highway Commission. *Highway Bulletin.* 1920–1925.

Oregon. *General Laws.* 1913–1940.

Oregon. Legislature. House. *Isaac Lee Patterson Bridge.* 36th Assembly. House Concurrent Resolution No. 1, 24 February 1931.

Oregon. Oregon State College. *Annual Catalogue.* 1912–1921.

Oregon. Oregon State College. *President's Biennial Report.* 1912–1916.

Oregon. Salem. Long Range Planning Commission. *A Long Range Plan for Salem, Oregon: First Annual Progress Report.* Salem: 1947.

Oregon. State Highway Commission. *Annual* and *Biennial Report.* 1913–1948.

South Carolina. State Highway Commission. *Annual Report.* 1921–1933.

Texas. State Highway Department. *Texas Highway Bulletin.* 1924–1929.

U.S. Department of Agriculture. *Annual Reports.* 1900–1940 [Report of the Secretary and Report of the Chief of the Office of Public Road Inquiry, Office of Public Roads, and after 1916 the Bureau of Public Roads].

U.S. Patent Office. *Official Gazette of the United States Patent Office.* Vol. 181 (6 August 1912). patent no. 1,035,026 [James B. Marsh patentee].

U.S. Department of Transportation. Federal Highway Administration. *America's Highways, 1776-1976: A History of the Federal Aid Program.* Washington: GPO, [1977].

U.S. Works Progress Administration. *Oregon: End of the Trail,* revised ed. Portland, [Binsford and Mort], 1940.

Utah. State Road Commission. *Biennial Report.* 1913–1940.

Washington. State Highway Commission. *Biennial Report.* 1910–1940.

State Bridge Inventories

Atkins, Stephen B., and William E. Keeler. *Survey of Metal Truss, Swing and Vertical Lift Bridges in Florida.* Tallahassee: Florida Department of Transportation, Bureau of Environment, Environmental Research, 1981.

Clouette, Bruce, and Matthew Roth. *Connecticut's Historic Highway Bridges.* [Hartford]: Connecticut Department of Transportation, 1991.

————, and ————. *Rhode Island Historic Bridge Inventory, Part I: Inventory and Recommendations, Part II: National Register Form.* Providence: Rhode Island Department of Transportation, 1988.

Coburn, Gary, ed. *The Ohio Historic Bridge Inventory, Evaluation and Preservation Plan.* Columbus: Ohio Department of Transportation in cooperation with the Federal Highway Administration, 1983.

Cooper, James L. *Iron Monuments to Distant Posterity: Indiana's Metal Bridges, 1870-1930.* Greencastle: DePauw University, Federal Highway Administration, Indiana Department of Highways, Indiana Department of Natural Resources, and National Park Service, 1987.

Deibler, Dan Grove, and Paula A. C. Spero. *Metal Truss Bridges in Virginia: A Survey and Photographic Inventory of Metal Truss Bridges in Virginia, 1865-1932.* Charlottesville:

Virginia Highway and Transportation Research Council, May 1975–June 1982 [reports 1–5 by Deibler, reports 6–9 by Spero].

Design, Fraser. *Arizona Bridge Inventory: A Historical Inventory for the Arizona Department of Transportation*. Phoenix: Arizona Department of Highways, 1987.

———. *Colorado Bridge Survey: An Inventory for the Colorado Department of Highways*. Denver: Colorado Department of Highways, 1984.

———. *Wyoming Truss Bridge Survey*. Cheyenne: Wyoming State Highway Department, 1982.

Elling, Rudolf E., and Gaylord B. Witherspoon. *Metal Truss Highway Bridge Inventory for the South Carolina Department of Highways and Public Transportation*. Clemson: Clemson University Press, 1981.

Fore, George. *North Carolina's Metal Truss Bridges: An Inventory and Evaluation*. Raleigh: North Carolina Division of Archives and History, Department of Cultural Resources, and the North Carolina Division of Highways, Department of Transportation, 1979.

Frame, Robert M., III. *Historic Bridge Project: A Report to the State Historic Preservation Office of the Minnesota Historical Society and the Minnesota Department of Transportation*. St. Paul: Minnesota Department of Transportation, 1985.

Goldfarb, Stephen J. *Georgia Department of Transportation and Georgia Department of Natural Resources Historic Bridge Survey*. Atlanta: Georgia Department of Transportation, 1981.

Herbst, Rebecca. *Idaho Bridge Inventory, Volume 1: History*. Boise: Idaho Transportation Department, 1983.

———, and Vicki Rottman, eds., *Historic Bridges of Colorado*. Denver: Colorado Department of Highways, 1986 [adapted from inventory by Fraser Design].

Hess, Jeffrey A. *Final Report of the Minnesota Historic Bridge Survey: Part I and Part II*. 2 volumes. St. Paul: Minnesota Historical Society, 1988.

———, and Robert M. Frame, III. *Historic Highway Bridges in Wisconsin, Volume I: Stone and Concrete Arch Bridges, Volume II: Intensive Survey Forms*. Madison: Wisconsin Department of Transportation, 1986.

Hyde, Charles K. *Michigan's Highway Bridges: History and Assessment*. Ann Arbor: Bureau of History, Michigan Department of State and the Michigan Department of Transportation, 1985.

Kemp, Emory L. *West Virginia's Historic Bridges*. Charleston: West Virginia Department of Culture and History, West Virginia Department of Highways, and Federal Highway Administration, 1984.

King, Joseph E. "A Historical Overview of Texas Transportation, Emphasizing Roads and Bridges." Austin: State Department of Highways and Public Transportation, 1988.

McClurkan, Burney B. *Arkansas Historic Bridge Inventory, Evaluation Procedures and Preservation Plan*. Little Rock: Arkansas Highway and Transportation Department, Environmental Division, 1987.

McCormick, Taylor and Associates, Inc., Consulting Engineers, Philadelphia. *Historic Highway Bridges in Pennsylvania*. Harrisburg: Pennsylvania Historical and Museum Commission, and the Pennsylvania Department of Transportation, 1986 [*Preservation Guide* published in 1987].

Mikesell, Stephen D. *Historic Highway Bridges of California*. Sacramento: California Department of Transportation [CalTrans], 1990.

Norman, James. *Oregon Covered Bridges: A Study for the 1989-90 Legislature*. Salem: Oregon Department of Transportation, Highway Division, Environmental Section, 1988.

Quivik, Fredric L. *Historic Bridges in Montana*. Denver: Rocky Mountain Region, National Park Service, 1982. [cosponsored by the State of Montana Department of Highways in cooperation with the U.S. Department of Transportation, Federal Highway Administration]

————, and Lon Johnson. *Historic Bridges of South Dakota*. [Pierre]: South Dakota Department of Transportation, 1990.

Rae, Steven R., Joseph E. King, and Donald R. Abbe. *New Mexico Historic Bridge Survey*. Santa Fe: New Mexico State Highway and Transportation Department and the Federal Highway Administration, Region 6, 1987.

Rawlings, G. D. *A Survey of Truss, Suspension and Arch Bridges in Kentucky*. Frankfort: Kentucky Department of Transportation, Bureau of Highways, Division of Environmental Analysis, 1982.

Smith, Dwight A., James B. Norman, and Pieter T. Dykman. *Historic Highway Bridges of Oregon*. 2d ed., revised. Portland: Oregon Historical Society Press, 1989 [original edition issued by Oregon Department of Transportation, 1985].

Spero, P. A. C., and Co. *Delaware Historic Bridges Survey and Evaluation*. Dover: Delaware Department of Transportation, Division of Highways, 1991.

Spero, Paula A. C. *A Survey and Photographic Inventory of Concrete and Masonry Arch Bridges in Virginia*. Charlottesville: Virginia Highway and Transportation Research Council, 1984.

Index

Jacob Conser Bridge (Marion and Linn counties), 138
John McLoughlin Bridge (Clackamas County): American Institute of Steel Construction excellence in design award, 97, 98, 99, 109, 128, 167 (n44); description of, 138; design of, 97, 99; photographs of, 98

Kelley, Earl Foster, 31, 34, 36, 126
Kerr, William Jasper, 38, 43, 97
Keystone Bridge Company, 2
Kiewit Construction, 133
King Bridge Company, 2, 18, 19
Kinne, W.S., 64
Kirkham, John Edward, 3, 16, 17–18, 32, 145 (nn28–29), 149 (n46)
Klamath River (Keno) Bridge (Klamath County), 137
Klein, Roy A., 68, 78

Lancaster, Samuel C., 40
Landmark American Bridges (DeLony), 134
Lane, Wallace R., 33
Latourell Creek Bridge (Multnomah County), 51; drawing of, 45
LAW, 21
Law: engineering and, convergence of, 4, 33–34, 65–67, 123, 127, 148 (n38), 149 (n40), 172 (n37); McCullough's education in, 4, 67, 78, 157–58 (n26)
League of American Wheelmen (LAW), 21
Lewis, John H., 38, 151 (n12[40])
Lewis and Clark River Bridge (Clatsop County), 141; photograph of, 104
Line of pressure bridge design method, 81
Long Range Plan for Salem, Oregon: First Annual Progress Report, A, 125
Luten, Daniel B., 16, 25, 33, 34, 148 (n38), 149 (n41)
Luten Engineering Company, 33, 145–46 (n37)
Luten v. Marsh Engineering Company, 33–34, 126, 148 (n38), 149 (n40)

MacDonald, Thomas H.: as Bureau of Public Roads head, 3, 35–36, 46–47, 153 (n28); education of, 146 (n6); Federal-Aid Road Act and, 47, 58–59; highway beautification and, 167 (n47[101]); Highway Research Board and, 64; Inter-American Highway and, 109, 111, 112, 124; at Iowa State Highway Commission, 3, 22–25, 29, 33, 126, 146–47 (nn7–8); Isaac Lee Patterson Bridge and, 72; toll bridge opposition of, 96
Maillart, Robert, 2, 134
Marsh, James Barney, 3, 18–19, 32, 83, 126, 145–46 (nn36–38)
Marsh Bridge Company, 18. *See also* Marsh Engineering Company
Marsh Engineering Company, 3, 18, 24
Marsh Engineering Company, Luten v., 33–34, 126, 148 (n38), 149 (n40)
Marston, Anson: civil engineering curriculum of, 9–10, 13–15, 17, 144 (n21); death of, 149 (n46); education of, 10, 11, 12, 127; on essential qualities of engineers, 43–44, 128; Illinois Central Railroad employment of, 12; on intelligence of McCullough, 20; Iowa Agricultural College career of, 9, 12, 13–14, 149 (n46); Iowa State Highway Commission and, 22, 23, 24–25, 27, 28, 149 (n46); as mentor of McCullough, 3, 6–7, 126, 127; Missouri Pacific Railroad employment of, 12; photograph of, 10; William Jasper Kerr and, 38
Mathematical Theory of Vibration in Suspension Bridges, The, 121
McCullough, Boyd (grandfather), 7
McCullough, Conde Balcom:
Awards and honors: Coos Bay Bridge dedicated as Conde B. McCullough Memorial Bridge, 125; Engineering Hall of Fame (Oregon State University), 134; *Engineers of Dreams: Great Bridge Builders and the Spanning of America,* recognition by, 125; *ENR*